INTERNATIONAL CENTRE FOR MECHANICAL SCIENCES

COURSES AND LECTURES No. 119

IAN N. SNEDDON
UNIVERSITY OF GLASGOW

THE LINEAR THEORY OF THERMOELASTICITY

COURSE HELD AT THE DEPARTMENT
OF MECHANICS OF SOLIDS
JULY 1972

UDINE 1974

SPRINGER-VERLAG WIEN GMBH

Originally published by Springer-Verlag Wien-New York in 1972

ISBN 978-3-211-81257-0 ISBN 978-3-7091-2648-6 (eBook)

DOI 10.1007/978-3-7091-2648-6

PREFACE

This monograph is made up of the notes for the lectures I gave at the International Centre for Mechanical Sciences in Udine in June 1972.

Chapter I contains a brief account of the derivation of the basic equations of the linear theory of thermoelasticity. The propagation of thermoelastic waves is considered in some detail in Chapter 2, while boundary-value and initial value problems are considered in the remaining two chapters. In Chapter 3 an account is given of coupled problems of thermoelasticity, and in the final chapter the methods traditionally used in the solution of static problems in thermoelasticity are outlined. Since integral transform methods are used where appropriate, the main properties of the most frequently used integral transforms are collected together in an Appendix.

It is my pleasant duty to record here my thanks to the officers of C.I.S.M. and in particular to Professor Luigi Sobrero and Professor Woclaw Olszak for inviting me to give the lectures and for making my stay in Udine so very enjoyable and stimulating.

Udine, June 1972

Ian N. Sneddon

CHAPTER I

THE BASIC EQUATIONS OF THE LINEAR THEORY OF THERMOELASTICITY

1.1 Introduction

The theory of thermoelasticity is concerned with the effect which the temperature field in an elastic solid has upon the stress field in that solid and with the associated effect which a stress field has upon thermal conditions in the solid. In the "classical" theory attention is restricted to perfectly elastic solids undergoing infinitesimal strains and infinitesimal fluctuations in temperature. It turns out that these assumptions are sufficiently strong to justify a purely thermodynamic derivation of the *form* of the equation of state of an elastic solid. The resulting equations are linear only if the range of fluctuation of temperature is such that the variation with temperature of the mechanical and thermal constants of the solid may be neglected.

In these notes we shall also make the basic assumption that there exists a state in which all of the components of strain, stress and temperature gradient vanish identically.

We shall assume, in addition, that the solid is *isotropic*. This is not an essential assumption. It is simply one which makes the equations much easier to handle. The corresponding equations for an isotropic body are still linear - but a good deal more complicated than those derived here.

The study of thermal stresses was begun by Duhamel (1838) who derived the additional terms which must be added to the components of the strain tensor when a temperature gradient has been set up in the body. Duhamel's results were rediscovered by Neumann (1885) who applied them in a study of the double refracting property of unequally heated glass plates. Neumann also showed that these additional terms in the strain components can be interpreted as meaning that the effect of a non-uniform distribution throughout an elastic solid is equivalent to that of a body force which is proportional to the temperature gradient.

The defect of the Duhamel-Neumann theory is that although it predicts the effect of a non-uniform temperature field on the strain field, it leads to the conclusion that the process of the conduction of heat through an elastic solid is not affected by the deformation of the solid. Both Duhamel and Neumann were conscious of this defect in their theory and (again, independently) suggested on purely empirical grounds that a term proportional to the time rate of change of the dilatation should be added to the heat-conduction equation. A number of authors notably Voigt (1910), Jeffreys (1930), Lessen and Duke (1953) and Lessen (1956) adduced thermodynamic arguments to justify the coupled equations postulated by Duhamel and Neumann. The failure of these authors to establish a satisfactory basis for thermoelasticity stems from their use of classical (i.e. *reversible*) thermodynamics. For, although the deformation of a perfectly elastic solid is a reversible process, the diffusion of heat occurs irreversible so that the derivation of the field equations has to be based on the theory of irreversible processes. The development of a satisfactory theory of irreversible thermodynamics - an excellent account of which is given in de Groot (1952) - provided the tools necessary to establish a proper theory of thermoelasticity. This theory was established by Biot (1956).

In these lectures we shall be concerned more with the solution of the equations of thermoelasticity than with a full discussion of their derivation. In this first chapter, however, a *brief sketch* is given of the derivation of the basic field equations and of the variational principles which are their equivalent. Full accounts are given by Nowacki (1962) and Kovalenko (1969) and a more brief outline by Chadwick (1960). The present discussion leans heavily on the last two works quoted.

To facilitate the solution of problems in cylindrical and spherical polar coordinates the equations are derived in terms of these coordinates in § 8.

Throughout we have used the notation of Green and Zerna (1954) for the stress tensor, denoting stress components by σ_{ij} $(i,j = 1,2,3)$ or in particular problems by $\sigma_{xx}, \sigma_{yy}, \sigma_{zz}, \sigma_{yz}, \sigma_{xz}, \sigma_{xy}$. For the components of the strain tensor we use

ε_{ij} so that for an infinitesimal strain,

$$\varepsilon_{ij} = \frac{1}{2}\left(\frac{\partial u_j}{\partial x_i} + \frac{\partial u_i}{\partial x_j}\right)$$

$(i,j = 1,2,3)$.

1.2 The Basic Equations of Thermoelasticity

We consider a perfectly elastic solid, initially unstrained, unstressed and at a uniform temperature T_0. When the solid is deformed by either mechanical or thermal means a displacement field u and a non-uniform temperature field T are set up. These in turn lead to a velocity field $\vec{v}$ and a distribution of strain and stress described respectively by the strain tensor ε_{ij} and the stress tensor σ_{ij}. Following such a disturbance of the equilibrium state, energy is transferred from one part of the solid to another by the elastic deformation and by the conduction of heat. The deformation of a perfectly elastic solid is a reversible process but heat conduction takes place irreversibly so the derivation of the partial differential equations satisfied by the field quantities has to be based on the thermodynamics of irreversible processes.

If σ denotes the density of the solid in the initial state and (v_1, v_2, v_3) the components of the velocity vector $\vec{v}$ the *conservation of mass* is expressed by the equation

$$\frac{D\varrho}{Dt} + \frac{\partial}{\partial x_j}(\varrho v_j) = 0 \qquad (1.2.1)$$

where (x_1, x_2, x_3) are the coordinates of a typical field point and

$$\frac{D}{Dt} = \frac{\partial}{\partial t} + v_j \frac{\partial}{\partial x_j}$$

is the operator of convective time differentiation.

Similarly, the *conservation of linear momentum* is expressed by the three equations (*)

$$\varrho \frac{Dv_i}{Dt} = \varrho X_i + \frac{\partial \sigma_{ij}}{\partial x_j} , \quad (i = 1,2,3) \qquad (1.2.2)$$

where (X_1, X_2, X_3) are the components of the body force per unit mass.

If we denote by Q the rate at which heat is generated (per unit volume) by internal sources and introduce the heat flux vector $\vec{q} = (q_1, q_2, q_3)$ the equation expressing the *conservation of energy* is

$$\frac{D}{Dt}\left(U + \frac{1}{2}\varrho v_i v_i\right) = \varrho X_i v_i + \frac{\partial}{\partial x_j}(v_i \sigma_{ij}) - \frac{\partial q_i}{\partial x_i} + Q . \qquad (1.2.3)$$

(*) The summation convention is used consistently throughout these notes

The final equation of the set expresses *the second law of thermodynamics*. If U denotes the specific (*) internal energy and S the specific entropy of the system we have the equation

$$\frac{DU}{Dt} = T\frac{DS}{Dt} + \sigma_{ij}\frac{D\varepsilon_{ij}}{Dt} \,. \qquad (1.2.4)$$

If we form the scalar produce of both sides of the vector equation (1.2.2) with the velocity vector $\vec{v}$ we find that

$$\sigma\frac{D}{Dt}\left(\frac{1}{2}v_i v_i\right) = \sigma X_i v_i + v_i\frac{\partial\sigma_{ij}}{\partial x_j}$$

and substracting this equation from (1.2.3) that

$$\frac{DU}{Dt} = \sigma_{ij}\frac{\partial v_i}{\partial x_j} - \frac{\partial q_i}{\partial x_i} + Q \,.$$

From the symmetry of the stress tensor we deduce that

$$\sigma_{ij}\frac{\partial v_i}{\partial x_j} = \frac{1}{2}\sigma_{ij}\left(\frac{\partial v_i}{\partial x_j} + \frac{\partial v_j}{\partial x_i}\right)$$

so that this last equation may be written in the form

$$\frac{DU}{Dt} = \frac{1}{2}\sigma_{ij}\left(\frac{\partial v_i}{\partial x_j} + \frac{\partial v_j}{\partial x_i}\right) - \frac{\partial q_i}{\partial x_i} + Q \,.$$

(*) It should be observed that "*specific*" means "*per unit volume*" throughout.

1.3 The Equations of the Linear Theory

In the classical theory of thermoelasticity we assume that the disturbances from the equilibrium state are always small, i.e. that the deviations of all physical quantities are all so small that their products and their spatial derivatives can be consistently neglected. In particular we assume that

$$\max|T - T_0| \ll T_0 \tag{1.3.1}$$

and that the components of the strain tensor are given by the equations

$$\varepsilon_{ij} = \frac{1}{2}\left(\frac{\partial u_i}{\partial x_j} + \frac{\partial u_j}{\partial x_i}\right) \tag{1.3.2}$$

and that the convective differential operator D/Dt may be replaced by the simple operator $\partial/\partial t$.

We also assume that each of the elastic and thermal constants entering into the theory do not depend upon the state variables (strain and temperature).

In this approximation, equation (1.2.1) reduces to

$$\frac{1}{\sigma}\frac{\partial \sigma}{\partial t} + \vartheta = 0 \,,$$

where ϑ denotes the dilatation

$$\vartheta = \varepsilon_{ii} \,,$$

and this is easily seen to have solution

$$\rho = \rho_0 e^{-\vartheta}$$

where ρ_0 is the initial density of the solid (assumed to be uniform). Since ϑ is itself assumed to be small we see that this is equivalent to the equation

$$\rho = \rho_0(1 - \vartheta) . \tag{1.3.3}$$

Equation (1.2.2) reduces to the familiar form

$$\frac{\partial \sigma_{ij}}{\partial x_j} + \rho X_i = \rho \frac{\partial^2 u_i}{\partial t^2} \tag{1.3.4}$$

and equations (1.2.4) and (1.2.5) respectively to the forms

$$\frac{\partial U}{\partial t} = T \frac{\partial S}{\partial t} + \sigma_{ij} \frac{\partial \varepsilon_{ij}}{\partial t} , \tag{1.3.5}$$

$$\frac{\partial U}{\partial t} = \sigma_{ij} \frac{\partial \varepsilon_{ij}}{\partial t} - \frac{\partial q_i}{\partial x_i} + Q . \tag{1.3.6}$$

Subtracting these equations we obtain the equation

$$T \frac{\partial S}{\partial t} + \frac{\partial q_i}{\partial x_i} = Q \tag{1.3.7}$$

which can readily be put into the alternative form

$$\frac{\partial S}{\partial t} + \frac{\partial}{\partial x_i}\left(\frac{q_i}{T}\right) = \frac{1}{T}\left(Q - \frac{q_i}{T} \frac{\partial T}{\partial x_i}\right) . \tag{1.3.8}$$

Equation (1.3.8) expresses the balance of entropy in the system. If we integrate both sides of this equation over an arbitrary volume we see that

$$S = \frac{1}{T}\left(Q - \frac{q_i}{T}\frac{\partial T}{\partial x_i}\right) \qquad (1.3.9)$$

is the rate at which entropy is created per unit volume by the diffusion of heat through the medium. Using this result and the Onsager reciprocal relations we can deduce that there exist constants L_{ij} with $L_{ij} = L_{ji}$ such that

$$q_i = -\frac{L_{ij}}{T^2}\frac{\partial T}{\partial x_j} .$$

This is a form of *the Fourier Law of heat conduction.* Since we shall be concerned only with isotropic media we may take

$$L_{ij} = kT^2 \delta_{ij}$$

where k is the thermal conductivity, in which case the equation becomes

$$q_i = -k\frac{\partial T}{\partial x_i} \qquad (1.3.10)$$

Substituting for q_i from equation (1.3.10) into equation (1.3.7) we obtain the equation

$$T\frac{\partial S}{\partial t} = k\Delta_3 T + Q \qquad (1.3.11)$$

in which

$$\Delta_3 = \frac{\partial^2}{\partial x_i \partial x_i}$$

is the Laplacian operator in three dimensional space.

1.4 Thermodynamic Relations

The equations (1.3.4) and (1.3.11) are (within the classical theory) the field equations of a perfect elastic solid which conducts heat. We have still, however, to formulate the relations which determine the stress components σ_{ij} and the specific entropy S as functions of the displacement components u_i and the temperature T.

It is *not* possible within the framework of thermodynamics alone to derive the equation of state of an elastic solid which connects the stress components σ_{ij}, the strain components ε_{ij} and the temperature T, and upon a knowledge of which the derivation of the stress-strain relation is dependent. However the *form* of such an equation of state can be predicted provided we make use of such supplementary information as is available. The assumptions of the classical theory of elasticity are sufficiently strong for such a derivation to be possible, since, in the kind of solid body in which we are interested, the distribution of stress depends only upon the state variables ε_{ij}, T and not upon their time derivatives.

We recall that the *Helmholtz free energy* F is defined by the equation

$$F = U - TS \qquad (1.4.1)$$

Writing equation (1.3.5) in the differential form

$$dU - TdS = \sigma_{ij}\, d\varepsilon_{ij} \tag{1.4.2}$$

we see that equation (1.4.1) is equivalent to the equation

$$dF = -SdT + \sigma_{ij}\, d\varepsilon_{ij}$$

from which we deduce immediately that

$$\sigma_{ij} = \left(\frac{\partial F}{\partial \varepsilon_{ij}}\right)_T . \tag{1.4.3}$$

If we now expand the free energy F in a Taylor series in the variables ε_{ij}, $T - T_0$, from the isotropy property we can show that the equation connecting σ_{ij} with ε_{ij} and $T - T_0$ is of the form

$$\sigma_{ij} = \frac{E_T}{(1+\eta_T)(1-2\eta_T)}\left[\eta_T \vartheta \delta_{ij} + (1-2\eta_T)\varepsilon_{ij} - (1+\eta_T)\alpha(T-T_0)\delta_{ij}\right] . \tag{1.4.4}$$

In this equation E_T, η_T are respectively the Young's modulus and Poisson's ratio of the solid and, as the suffix T suggests, are the *isothermal elastic constants.*

If we contract this equation we find that $p = -\frac{1}{3}\sigma_{ii}$ is given by

$$p = \frac{-E_T}{3(1-2\eta_T)}\left[\vartheta - 3(T-T_0)\alpha\right] .$$

From this equation we see immediately that α is *the coefficient of linear expansion* (at constant stress) and that $\varkappa_T$, *the isothermal compressibility*

$$\varkappa_T = -\left(\frac{\partial \vartheta}{\partial p}\right)_T$$

has the value

$$\varkappa_T = \frac{3(1-2\eta_T)}{E_T} \tag{1.4.5}$$

for such a solid.

From equation (1.4.2) we have

$$dS = \frac{dU}{T} - \frac{1}{T}\sigma_{ij}\,d\varepsilon_{ij}$$

$$= \frac{1}{T}\left[\left(\frac{\partial U}{\partial T}\right)_{\varepsilon} dT + \left(\frac{\partial U}{\partial \varepsilon_{ij}}\right)_T d\varepsilon_{ij}\right] - \frac{1}{T}\sigma_{ij}\,d\varepsilon_{ij}$$

from which we deduce that

$$T\left(\frac{\partial S}{\partial T}\right)_{\varepsilon} = \left(\frac{\partial U}{\partial T}\right)_{\varepsilon} . \tag{1.4.6}$$

From the physical interpretation of the expression on the left-hand side of this equation we see that it is the *specific heat at constant strain* which we shall denote by c_ε. We therefore have the relation

$$\left(\frac{\partial U}{\partial T}\right)_{\varepsilon} = c_\varepsilon . \tag{1.4.7}$$

We also have the relation

$$\left(\frac{\partial S}{\partial \varepsilon_{ij}}\right)_T = \frac{1}{T}\left(\frac{\partial U}{\partial \varepsilon_{ij}}\right)_T - \frac{1}{T}\sigma_{ij} \ . \qquad (1.4.8)$$

From equation (1.4.1) we have

$$\left(\frac{\partial F}{\partial \varepsilon_{ij}}\right)_T = \left(\frac{\partial U}{\partial \varepsilon_{ij}}\right)_T - T\left(\frac{\partial S}{\partial \varepsilon_{ij}}\right)_T$$

$$\left(\frac{\partial F}{\partial T}\right)_\varepsilon = \left(\frac{\partial U}{\partial T}\right)_\varepsilon - T\left(\frac{\partial S}{\partial T}\right)_\varepsilon - S$$

and using equations (1.4.6) and (1.4.8) we see that these are equivalent to the pair of equations

$$\left(\frac{\partial F}{\partial \varepsilon_{ij}}\right)_T = \sigma_{ij} \ , \qquad \left(\frac{\partial F}{\partial T}\right)_\varepsilon = -S$$

from which we deduce that since, by equation (1.3.3),

$$\left(\frac{\partial \sigma}{\partial T}\right)_\varepsilon = 0 \ ,$$

then

$$\left(\frac{\partial S}{\partial \varepsilon_{ij}}\right)_T = -\left(\frac{\partial \sigma_{ij}}{\partial T}\right)_\varepsilon .$$

From equations (1.4.8) and (1.4.9) we then have the equation

$$\left(\frac{\partial U}{\partial \varepsilon_{ij}}\right)_T = \sigma_{ij} - T\left(\frac{\partial \sigma_{ij}}{\partial T}\right)_\varepsilon .$$

Substituting for σ_{ij} from equation (1.4.4), we find that

$$\left(\frac{\partial S}{\partial \varepsilon_{ij}}\right)_T = \frac{E_T \alpha}{(1-2\eta_T)} \delta_{ij} \tag{1.4.9}$$

$$\left(\frac{\partial U}{\partial \varepsilon_{ij}}\right)_T = \frac{E_T}{(1+\eta_T)(1-2\eta_T)} \left[\eta_T \vartheta \delta_{ij} + (1-2\eta_T)\varepsilon_{ij} + (1+\eta_T)\alpha T_0 \delta_{ij}\right] \tag{1.4.10}$$

Substituting from equations (1.4.7) and (1.4.10) into the formula

$$dU = \frac{\partial U}{\partial T} dT + \frac{\partial U}{\partial \varepsilon_{ij}} d\varepsilon_{ij}$$

and integrating we find that

$$U - U_0 = \frac{1}{2}\sigma_{ij}\varepsilon_{ij} + \frac{\alpha E_T}{2(1-2\eta_T)}(T+T_0)\vartheta + c_\varepsilon(T-T_0) \tag{1.4.11}$$

where U_0 denotes the value of U in the initial state $\varepsilon_{ij}=0, T=T_0$. The term $\frac{1}{2}\sigma_{ij}\,\varepsilon_{ij}$ is the elastic strain energy per unit volume of the body and $c_\varepsilon(T-T_0)$ is the heat content per unit mass, the remaining term

$$\frac{\alpha E_T}{2(1-2\eta_T)}(T+T_0)$$

is the consequence of the interaction of the elastic deformation and the thermal diffusion.

Similarly, from equations (1.4.6), (1.4.7) and (1.4.9) we deduce the equation

$$S-S_0 = \frac{\alpha E_T}{(1-2\eta_T)}\vartheta + c_\varepsilon \ln(T/T_0)\,. \qquad (1.4.12)$$

Since purely elastic changes take place reversibly they contribute nothing to the entropy change. The term $c_\varepsilon \ln(T/T_0)$ is the entropy change due to the heat conduction alone and the term

$$\frac{\alpha E_T}{(1-2\eta_T)}$$

arises from the coupling of the elastic and thermal changes.

When studying the thermodynamics of deformation it is sometimes also found advantageous to introduce, in addition to the specific free energy F , the *Gibbs free energy* G defined by the equation

$$G = F - \sigma_{ij}\varepsilon_{ij}\,.$$

The function G is also called the *Gibbs thermodynamic potential*. Since

$$dF = -SdT + \sigma_{ij}d\varepsilon_{ij} \qquad (1.4.13)$$

we deduce that

$$dG = -SdT - \varepsilon_{ij}d\sigma_{ij} \qquad (1.4.14)$$

and hence that

$$\left(\frac{\partial G}{\partial T}\right)_{\sigma} = -S\ . \quad \left(\frac{\partial G}{\partial \sigma_{ij}}\right)_T = -\varepsilon_{ij}\ . \qquad (1.4.15)$$

1.5 The Equations for the Temperature and Displacement Fields

From equation (1.4.12) we find that

$$\frac{\partial S}{\partial t} = \frac{\alpha E_T}{(1-2\eta_T)}\frac{\partial}{\partial t} + \frac{c_\varepsilon}{T}\frac{\partial T}{\partial t}\ .$$

If we substitute this expression into equation (1.3.11) and, by making the assumption (1.3.1), replace

$$T\frac{\partial \vartheta}{\partial t}$$

by

$$T_0\frac{\partial \vartheta}{\partial t}$$

and drop the suffix T we find that equation (1.3.11) becomes

$$\sigma c_\varepsilon \frac{\partial T}{\partial t} + \frac{\alpha E T_0}{3(1-2\eta)}\frac{\partial \vartheta}{\partial t} = k\Delta_3 T + Q\ . \qquad (1.5.1)$$

Similarly, if we substitute the expression (1.4.4) for σ_{ij} into equation (1.3.4) we find it becomes

$$\eta\frac{\partial \vartheta}{\partial x_i} + (1-2\eta)\frac{\partial \varepsilon_{ij}}{\partial x_j} - (1+\eta)\alpha\frac{\partial T}{\partial x_i} =$$

$$= \frac{(1+\eta)(1-2\eta)\sigma}{E}\left(\frac{\partial^2 u_i}{\partial t^2} - X_i\right). \tag{1.5.2}$$

These equations can easily be written in vector form. If $\vec{u} = (u_1, u_2, u_3)$, then

$$\frac{\partial \varepsilon_{ij}}{\partial x_j} = \frac{1}{2}\left[\Delta_3 \vec{u} + \text{grad div } \vec{u}\right], \quad \frac{\partial \vartheta}{\partial x_i} = \text{grad div } \vec{u}$$

so that equation (1.5.2) can be rewritten as

$$(1-2\eta)\Delta_3 \vec{u} + \text{grad}(\text{div } \vec{u}) - 2(1+\eta)\alpha \, \text{grad } T =$$

$$= \frac{2(1+\eta)(1-2\eta)\sigma}{E} \frac{\partial^2 \vec{u}}{\partial t^2} - X. \tag{1.5.3}$$

In the same notation (1.5.1) is

$$\sigma c_\varepsilon \frac{\partial T}{\partial t} + \frac{\alpha E T_0}{(1-2\eta)} \frac{\partial}{\partial t} \text{div } v = k\Delta_3 T + Q. \tag{1.5.4}$$

If we express the displacement vector $\vec{u}$ as the sum of an irrotational vector and a solenoidal vector through the formula

$$\vec{u} = \text{grad } \phi + \text{curl } \vec{A} \tag{1.5.5}$$

and make use of

$$\operatorname{div}\operatorname{curl}\vec{A} = 0\,,$$

$$\operatorname{curl}\operatorname{grad}\phi = 0\,,$$

$$\operatorname{curl}\operatorname{curl}\vec{A} = \operatorname{grad}\operatorname{div}\vec{A} - \Delta_3\vec{A}$$

we find that *in the absence of body forces and heat sources* $(X=0\,, Q=0)$,equations (1.5.3) and (1.5.4) become

$$\frac{\partial^2\phi}{\partial t^2} = \frac{(1-\eta)E}{\sigma(1+\eta)(1-2\eta)}\Delta_3\phi - \frac{\alpha E}{\sigma(1-2\eta)}(T-T_0) \tag{1.5.6}$$

$$k\Delta_3 T = \sigma c_\varepsilon \frac{\partial T}{\partial t} + \frac{\alpha E T_0}{(1-2\eta)}\frac{\partial}{\partial t}\Delta_3\phi \tag{1.5.7}$$

$$\frac{\partial^2\vec{A}}{\partial t^2} = \frac{E}{2\sigma(1+\eta)}\Delta_3\vec{A}\,. \tag{1.5.8}$$

Eliminating T from equations (1.5.6) and (1.5.7) we find that ϕ is a solution of the equation

$$\left(\frac{\partial}{\partial t} - \frac{k}{\sigma c_\varepsilon}\Delta_3\right)\left(\frac{\partial^2}{\partial t^2} - V_T^2\Delta_3\right)\phi = \frac{\alpha^2 T_0}{\sigma^2 c_\varepsilon x^2}\frac{\partial}{\partial t}\Delta_3\phi \tag{1.5.9}$$

where V_T , the velocity of dilatational waves in medium with zero coefficient of expansion (*), is defined the equation

(*) We shall refer to V_T as the *isothermal velocity.*

$$V_T^2 = \frac{(1-\eta)}{(1+\eta)(1-2\eta)}\left(\frac{E}{\sigma}\right) \qquad (1.5.10)$$

and $\varkappa$ denotes the isothermal compressibility defined by equation (1.4.5).

Many authors take λ and μ , the Lamé constants, as the basic elastic constants instead of E and η ; the relation between the two pairs of constant is given by the equations

$$\lambda = \frac{\eta E}{(1+\eta)(1-2\eta)}, \quad \mu = \frac{E}{2(1+\eta)} . \qquad (1.5.11)$$

In terms of the Lamé constants equations (1.4.4) and (1.5.3) may be rewritten respectively as

$$\sigma_{ij} = \lambda \vartheta \delta_{ij} + 2\mu \varepsilon_{ij} - (3\lambda + 2\mu)\alpha(T-T_0)\delta_{ij} , \qquad (1.5.12)$$

$$\mu \Delta_3 \vec{u} + (\lambda + \mu)\,\mathrm{grad}(\mathrm{div}\,\vec{u}) - (3\lambda + 2\mu)\alpha\,\mathrm{grad}\,T = \sigma \frac{\partial^2 \vec{u}}{\partial t^2} - \vec{X} . \qquad (1.5.13)$$

and equations (1.5.6), (1.5.7), (1.5.8) and (1.5.10) become

$$\frac{\partial^2 \phi}{\partial t^2} = V_T^2 \Delta_3 \phi - \frac{3\lambda - 2\mu}{\sigma}\alpha(T-T_0) , \qquad (1.5.14)$$

$$k\Delta_3 T = \sigma c_\varepsilon \frac{\partial T}{\partial t} + (3\lambda + 2\mu)\alpha T_0 \frac{\partial}{\partial t}\Delta_3 \phi , \qquad (1.5.15)$$

$$\frac{\partial^2 \vec{A}}{\partial t^2} = \frac{\mu}{\sigma}\Delta_3 \vec{A} , \qquad (1.5.16)$$

$$V_T^2 = \frac{\lambda + 2\mu}{\sigma} . \tag{1.5.17}$$

From these equations we see that the scalar function Φ describes *compressional waves* in which volume changes occur while the vector potential A describes *shear waves*, propagates with velocity $\sqrt{(\mu/\sigma)}$, which produce no volume changes.

1.6 Variational Principles

One of the standard methods in mathematical physics of obtaining approximate solutions of boundary value problems is to reformulate the problem as one in the calculus of variations. Biot (1956) has established such a variational principle for the coupled equations of elasticity.

It is convenient to introduce the concept of the *entropy flux* which is defined in terms of the heat flux s by the equation

$$T\dot{s} = \vec{q} . \tag{1.6.1}$$

Since in the linear theory we are making the assumption that all physical changes satisfy the requirement (1.3.1) we see that, in the linear theory, this defining equation is replaced by

$$T_0 \dot{s} = \vec{q} .$$

By allowing the components u_i of the displacement vector and the components s_i of the entropy flux to take independent variations δu_i , δs_i , Biot established the following variational principle for the coupled problem of thermoelasticity for a body occupying the region

$$\delta B + \delta D - \delta W = 0 \qquad (1.6.2)$$

where

$$B = \int_{\Omega} \left[\frac{1}{2}\lambda \vartheta^2 + \mu \varepsilon_{ij}\varepsilon_{ij} + \frac{1}{2}c_{\varepsilon}(T-T_0)^2/T_0\right] d\tau \qquad (1.6.3)$$

denotes what is called the *Biot thermoelastic potential*,

$$D = \frac{T_0}{2k}\int_{\Omega} \dot{s}_i \dot{s}_i d\tau \ , \quad (\dot{s}_i = q_i/T_0) \ , \qquad (1.6.4)$$

is called the *dissipation function*, and

$$\delta W = \int_{\partial\Omega} [f_i \delta u_i - (T-T_0) n_i \delta s_i] d\Sigma - \sigma \int_{\Omega} \ddot{u}_i \delta u_i d\tau \qquad (1.6.5)$$

denotes the variation in the *generalized virtual work*. In equation (1.6.5) $d\Sigma$ denotes the element of surface area and the integration is taken over the boundary $\partial\Omega$ of the region Ω, $\vec{n} = (n_1, n_2, n_3)$ is the unit outward drawn normal to $\partial\Omega$ and

$$f_i = \sigma_{ij} n_j \qquad (1.6.6)$$

is the applied surface stress.

In *uncoupled* thermoelastic problems, only the

components of displacement u_i are varied and the variational principle reduces (in the statical case) to the form

$$\delta\mathfrak{F} - \delta W = 0 \tag{1.6.7}$$

where $\mathfrak{F}$ is the integral over Ω of the specific free energy F ;

$$\mathfrak{F} = \int_{\Omega} F d\tau \tag{1.6.8}$$

and

$$W = \int_{\Omega} f_i u_i d\tau \ . \tag{1.6.9}$$

This result can, of course, easily be established directly from the equation $\sigma_{ij,j} = 0$ by means of the divergence theorem and the relations

$$\sigma_{ij} = \frac{\partial F}{\partial \varepsilon_{ij}}$$

(cf. Section 1.4 above).

Alternatively, equation (1.6.7) can be rewritten as

$$\delta\mathfrak{G} + \delta W = 0 \tag{1.6.10}$$

where $\mathfrak{G}$ is the integral over Ω of the Gibbs thermodynamic potential

$$\mathfrak{G} = \int_{\Omega} G d\tau \tag{1.6.11}$$

and W is again defined by equation (1.6.9).

Similarly, we obtain the variational equation for heat conduction by putting all of the mechanical terms equal to zero. Again it is of the form (1.6.2) but now

$$B = \frac{1}{2T_0}\int_{\Omega} c_{\varepsilon}(T-T_0)^2 d\tau \tag{1.6.12}$$

$$D = \frac{T_0}{2k}\int_{\Omega} \dot{s}_i \dot{s}_i d\tau\,, \quad (\dot{s}_i = \dot{q}_i/T_0)\,, \tag{1.6.13}$$

and

$$\delta W = -\int_{\partial\Omega} (T-T) n_i \delta s_i d\Sigma\,. \tag{1.6.14}$$

1.7 Problems in the Plane

In the classical theory of elasticity we distinguish between two types of plane problem:

(*a*) *plane strain* which is the kind of deformation which occurs in very long cylinders whose generators may be taken to be parallel to the z-axis.

(*b*) *plane stress* which is the state which is the kind of deformation which occurs in thin plates whose plane faces are free from stress.

We shall discuss the two cases separately.

Case (a) Plane Strain:

In this case we assume that the loading is such that the stress is the same in all normal cross-sections of the cylinder.

This is equivalent to assuming that the displacement vector is of the form

$$\vec{u} = \{u_x(x,y), u_y(x,y), 0\} \tag{1.7.1}$$

and hence from equation (1.4.4) that the stress strain relations become

$$\sigma_{xx} = \frac{E}{(1+\eta)(1-2\eta)}\left[(1-\eta)\frac{\partial u_x}{\partial x} + \frac{\partial u_y}{\partial y} - (1+\eta)\alpha(T-T_0)\right] \tag{1.7.2}$$

$$\sigma_{yy} = \frac{E}{(1+\eta)(1-2\eta)}\left[\eta\frac{\partial u_x}{\partial x} + (1-\eta)\frac{\partial u_y}{\partial y} - (1+\eta)\alpha(T-T_0)\right] \tag{1.7.3}$$

$$\sigma_{zz} = \frac{E}{(1+\eta)(1-2\eta)}\left[\eta\frac{\partial u_x}{\partial x} + \eta\frac{\partial u_y}{\partial y} - (1+\eta)\alpha(T-T_0)\right] \tag{1.7.4}$$

$$\sigma_{xy} = \frac{E}{2(1+\eta)}\left[\frac{\partial u_y}{\partial x} + \frac{\partial u_x}{\partial y}\right] \tag{1.7.5}$$

$$\sigma_{xz} = \sigma_{yz} = 0\ . \tag{1.7.6}$$

Eliminating $\partial u_x/\partial x$ and $\partial u_y/\partial y$ from equations (1.7.2), (1.7.3) and (1.7.4) we obtain the simple relation

$$\sigma_{zz} = \eta(\sigma_{xx} + \sigma_{yy}) - \alpha E(T - T_0) \, . \tag{1.7.7}$$

The components of strain can be expressed through the simple equations

$$\varepsilon_{xx} = \frac{1+\eta}{E}\left[(1-\eta)\sigma_{xx} - \eta\sigma_{yy} + \alpha(1+\eta)(T - T_0)\right] \tag{1.7.8}$$

$$\varepsilon_{yy} = \frac{1+\eta}{E}\left[(1-\eta)\sigma_{yy} - \eta\sigma_{xx} + \alpha(1+\eta)(T - T_0)\right] \tag{1.7.9}$$

$$\varepsilon_{xy} = \frac{1+\eta}{E}\sigma_{xy} \, . \tag{1.7.10}$$

We can then express the *compatibility condition*

$$\frac{\partial^2 \varepsilon_{xx}}{\partial y^2} + \frac{\partial^2 \varepsilon_{yy}}{\partial x^2} = 2\frac{\partial^2 \varepsilon_{xy}}{\partial x \partial y} \tag{1.7.11}$$

in the alternative form

$$(1-\eta)\Delta_2(\sigma_{xx} + \sigma_{yy}) - \frac{\partial}{\partial x}\left(\frac{\partial \sigma_{xx}}{\partial x} + \frac{\partial \sigma_{xy}}{\partial y}\right) - \frac{\partial}{\partial y}\left(\frac{\partial \sigma_{xy}}{\partial x} + \frac{\partial \sigma_{yy}}{\partial y}\right) + \alpha E \Delta_2 T = 0, \tag{1.7.12}$$

where

$$\Delta_2 = \frac{\partial^2}{\partial x^2} + \frac{\partial^2}{\partial y^2}$$

is the two-dimensional Laplace operator.

In the case of *equilibrium* the equations (1.3.4) reduce to

$$\frac{\partial \sigma_{xx}}{\partial x} + \frac{\partial \sigma_{xy}}{\partial y} + \sigma X_x = 0 \tag{1.7.13}$$

$$\frac{\partial \sigma_{xy}}{\partial x} + \frac{\partial \sigma_{yy}}{\partial y} + \sigma X_y = 0 \tag{1.7.14}$$

so that equation (1.7.12) becomes

$$(1-\eta)\Delta_2(\sigma_{xx}+\sigma_{yy}) + \sigma\left(\frac{\partial X_x}{\partial x} + \frac{\partial X_y}{\partial y}\right) + \alpha E \Delta_2 T = 0 . \tag{1.7.15}$$

If, in addition, the body force is conservative, i.e. if there exists a potential function $V(x,n)$ such that

$$X_x = -\frac{\partial V}{\partial x}, \quad X_y = -\frac{\partial V}{\partial y} \tag{1.7.16}$$

we see that equations (1.7.13) and (1.7.14) can be written as

$$\frac{\partial}{\partial x}(\sigma_{xx} - \sigma V) + \frac{\partial}{\partial y}\sigma_{xy} = 0 , \quad \frac{\partial}{\partial x}\sigma_{xy} + \frac{\partial}{\partial y}(\sigma_{yy} - \sigma V) = 0$$

and equation (1.7.15) as

$$(1-\eta)\Delta_2(\sigma_{xx}+\sigma_{yy})-\sigma\Delta_2 V+\alpha E\Delta_2 T = 0 .$$

It follows immediately that a solution of equations (1.7.13), (1.7.14) is given by the equations

$$\sigma_{xx} = \frac{\partial^2 \chi}{\partial y^2}+\sigma V , \quad \sigma_{xy} = -\frac{\partial^2 \chi}{\partial x \partial y} , \quad \sigma_{yy} = \frac{\partial^2 \chi}{\partial x^2}+\sigma V \quad (1.7.17)$$

where the function $\chi(x,y)$, which is called the *Airy stress function*, satisfies the fourth order partial differential equation

$$(1-\eta)\Delta_2^2\chi+(1-2\eta)\sigma\Delta_2 V+\alpha E\Delta_2 T = 0 \quad (1.7.18)$$

with

$$\Delta_2^2 = \Delta_2\Delta_2 = \frac{\partial^4}{\partial x^4}+2\frac{\partial^4}{\partial x^2\partial y^2}+\frac{\partial^4}{\partial y^4}$$

denoting the plane biharmonic operator.

Case (b) Plane Stress:

The state of plane stress occurs in a very thin plate whose central plane lies in the xy-plane and no forces are applied to the flat surface of the plate. It is assumed that

$$\sigma_{xz} = \sigma_{yz} = \sigma_{zz} = 0 . \quad (1.7.19)$$

The condition $\sigma_{zz}=0$ implies that

$$\varepsilon_{zz} = -\frac{\eta}{1-\eta}(\varepsilon_{xx}+\varepsilon_{yy})+\frac{1+\eta}{1-\eta}\alpha(T-T_0) .$$

Substituting this in equation (1.4.4) we find that

$$\sigma_{xx} = \frac{E}{(1+\eta)(1-\eta)}\varepsilon_{xx}+\eta\varepsilon_{yy}-(1+\eta)\alpha(T-T_0) .$$

If we make the substitutions

$$E = \frac{E_1}{1-\eta_1^2} , \quad \eta = \frac{\eta_1}{1-\eta_1} , \quad \alpha = \frac{\alpha_1}{1+\eta_1} \tag{1.7.20}$$

we see that this reduces to

$$\sigma_{xx} = \frac{E_1}{(1+\eta_1)(1-2\eta_1)}\left[(1-\eta_1)\varepsilon_{xx}+\nu_1\varepsilon_{yy}-(1+\eta_1)\alpha_1(T-T_0)\right] \tag{1.7.21}$$

and that we similarly obtain

$$\sigma_{yy} = \frac{E_1}{(1+\eta_1)(1-2\eta_1)}\left[\eta_1\varepsilon_{xx}+(1-\eta_1)\varepsilon_{yy}-(1+\eta_1)\alpha_1(T-T_0)\right] \tag{1.7.22}$$

$$\sigma_{xy} = \frac{E_1}{(1+\eta_1)}\varepsilon_{xy} . \tag{1.7.23}$$

Comparing this set of equations with equations (1.7.2), (1.7.3), (1.7.4) we see that as far as the calculation of the components of stress is concerned, the problem of plane stress can be reduced to that of plane strain by the simple changes in the values of E , η and α given by equations (1.7.20).

1.8 Curvilinear Coordinates

In the solution of many boundary value problems in thermoelasticity it is convenient to employ orthogonal curvilinear coordinates. We shall list here the commonly used formulae for cylindrical coordinates and spherical polar coordinates.

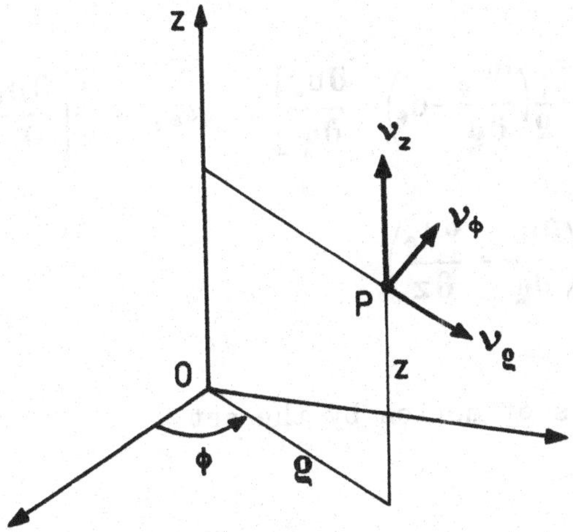

(*i*) *Cylindrical coordinates* (ϱ, ϕ, z)

$$\operatorname{grad} \Psi = \left(\frac{\partial \Psi}{\partial \varrho}, \frac{1}{\varrho}\frac{\partial \Psi}{\partial \phi}, \frac{\partial \Psi}{\partial z}\right)$$

$$\operatorname{div} \vec{v} = \frac{1}{\varrho}\frac{\partial}{\partial \varrho}(\varrho v_\varrho) + \frac{1}{\varrho}\frac{\partial v_\phi}{\partial \phi} + \frac{\partial v_z}{\partial z} ,$$

$$\Delta_3\Psi = \frac{\partial^2\Psi}{\partial\varrho^2} + \frac{1}{\varrho}\frac{\partial\Psi}{\partial\varrho} + \frac{1}{\varrho^2}\frac{\partial^2\Psi}{\partial\phi^2} + \frac{\partial^2\Psi}{\partial z^2} .$$

The components of the strain tensor are given by the set of equations

$$\varepsilon_{\varrho\varrho} = \frac{\partial u_\varrho}{\partial\varrho} , \quad \varepsilon_{\phi\phi} = \frac{1}{\varrho}\left(\frac{\partial\phi_\varrho}{\partial\phi} + u_\varrho\right) , \quad \varepsilon_{zz} = \frac{\partial u_z}{\partial z}$$

$$\varepsilon_{\varrho\phi} = \frac{1}{2}\left[\frac{1}{\varrho}\left(\frac{\partial u_\varrho}{\partial\varrho} - u_\phi\right) + \frac{\partial u_\phi}{\partial\varrho}\right] , \quad \varepsilon_{\phi z} = \frac{1}{2}\left[\frac{\partial u_\phi}{\partial z} + \frac{1}{\varrho}\frac{\partial u_z}{\partial\phi}\right]$$

$$\varepsilon_{\varrho z} = \frac{1}{2}\left(\frac{\partial u_z}{\partial\varrho} + \frac{\partial u_\varrho}{\partial z}\right)$$

and the equations of motion by the set

$$\frac{\partial\sigma_{\varrho\varrho}}{\partial\varrho} + \frac{1}{\varrho}\frac{\partial\sigma_{\varrho\phi}}{\partial\phi} + \frac{\partial\sigma_{\varrho z}}{\partial z} + \frac{\sigma_{\varrho\varrho} - \sigma_{\phi\phi}}{\varrho} + \sigma X_\varrho = \sigma\frac{\partial^2 u_\varrho}{\partial t^2} \quad (1.8.1)$$

$$\frac{\partial\sigma_{\varrho\phi}}{\partial\varrho} + \frac{1}{\varrho}\frac{\partial\sigma_{\phi\phi}}{\partial\phi} + \frac{\partial\sigma_{\phi z}}{z} + \frac{2\sigma_{\varrho\phi}}{\varrho} + \sigma X_\phi = \sigma\frac{\partial^2 u_\phi}{\partial t^2} \quad (1.8.2)$$

$$\frac{\partial\sigma_{\varrho z}}{\partial\varrho} + \frac{1}{\varrho}\frac{\partial\sigma_\phi}{\partial\phi} + \frac{\partial\sigma_{zz}}{\partial z} + \frac{\sigma_{\varrho z}}{\varrho} + \sigma X_z = \sigma\frac{\partial^2 u_z}{\partial t^2} . \quad (1.8.3)$$

If the stress and temperature fields are symmetric about the z-axis so that $u_\phi = 0$ and $\partial/\partial\phi \equiv 0$ the non-vanishing components of the stress tensor are given by the equations

$$\sigma_{\varrho\varrho} = \frac{E}{(1+\eta)(1-2\eta)}\left[(1-\eta)\frac{\partial u_\varrho}{\partial\varrho} + \eta\frac{u_\varrho}{\varrho} + \frac{\partial u_z}{\partial z} - (1+\eta)\alpha(T-T_0)\right] \tag{1.8.4}$$

$$\sigma_{\phi\phi} = \frac{E}{(1+\eta)(1-2\eta)}\left[\eta\frac{\partial u_\varrho}{\partial\varrho} + (1-\eta)\frac{u_\varrho}{\varrho} + \eta\frac{\partial u_z}{\partial z} - (1+\eta)\alpha(T-T_0)\right] \tag{1.8.5}$$

$$\sigma_{zz} = \frac{E}{(1+\eta)(1-2\eta)}\left[\eta\frac{\partial u_\varrho}{\partial\varrho} + \frac{u_\varrho}{\varrho} + (1-\eta)\frac{\partial u_z}{\partial z} - (1+\eta)\alpha(T-T_0)\right] \tag{1.8.6}$$

$$\sigma_{\varrho z} = \frac{E}{2(1+\eta)}\left[\frac{\partial u_\varrho}{\partial z} - \frac{\partial u_z}{\partial\varrho}\right] \tag{1.8.7}$$

and the equations of motion reduce to

$$\frac{\partial\sigma_{\varrho\varrho}}{\partial\varrho} + \frac{\partial\sigma_{\varrho z}}{\partial z} + \frac{\sigma_{\varrho\varrho} - \sigma_{\phi\phi}}{\varrho} + \sigma X_\varrho = \sigma\frac{\partial^2 u_\varrho}{\partial t^2} \tag{1.8.8}$$

$$\frac{\partial\sigma_{\varrho z}}{\partial\varrho} + \frac{\partial\sigma_{zz}}{\partial z} + \frac{\sigma_{\varrho z}}{\varrho} + \sigma X_z = \sigma\frac{\partial^2 u_z}{\partial t^2} \tag{1.8.9}$$

$$2(1-\eta)\mathcal{B}_1 u_\varrho + (1-2\eta)\frac{\partial^2 u}{\partial z^2} + \frac{\partial^2 u_z}{\partial\varrho\,\partial z} - 2(1+\eta)\alpha\frac{\partial T}{\partial\varrho} =$$

$$= \frac{2(1+\eta)(1-2\eta)}{E}\sigma\left(\frac{\partial^2 u_\varrho}{\partial t^2} - X_\varrho\right) \tag{1.8.10}$$

$$\frac{1}{\varrho}\frac{\partial}{\partial\varrho}\varrho\frac{\partial u_\varrho}{\partial z} + (1-2\eta)\mathcal{B}_0 u_z + 2(1-\eta)\frac{\partial^2 u_z}{\partial z^2} - 2(1+\eta)\alpha\frac{\partial T}{\partial z} =$$

$$= \frac{2(1+\eta)(1-2\eta)}{E}\sigma\left(\frac{\partial^2 u_z}{\partial t^2} - X_z\right) \tag{1.8.11}$$

where $\mathcal{B}_\nu$ denotes the *Bessel operator*

$$\mathcal{B}_\nu = \frac{\partial^2}{\partial\varrho^2} + \frac{1}{\varrho}\frac{\partial}{\partial\varrho} - \frac{\nu^2}{\varrho^2} \tag{1.8.12}$$

$$\sigma c_\varepsilon\frac{\partial T}{\partial t} + \frac{\alpha E T_0}{(1-2\eta)}\frac{\partial}{\partial t}\left[\frac{1}{\varrho}\frac{\partial}{\partial\varrho}(\varrho u_\varrho) + \frac{\partial u_z}{\partial z}\right] = k\left(\mathcal{B}_0 + \frac{\partial^2}{\partial z^2}\right)T + Q\ .$$

(ii) Spherical Polar Coordinates (r, θ, ϕ)

$$\text{grad}\,\Psi = \frac{\partial\Psi}{\partial r}, \frac{1}{r}\frac{\partial\Psi}{\partial\theta}, \frac{1}{r\sin\theta}\frac{\partial\Psi}{\partial\phi}$$

$$\text{div}\,\vec{v} = \frac{1}{r^2}\frac{\partial(r^2 v_r)}{\partial r} + \frac{1}{r\sin\theta}\frac{\partial}{\partial\theta}(\sin\theta\, v_\theta) + \frac{1}{r\sin\theta}\frac{\partial v_\phi}{\partial\phi}$$

$$\Delta_3\Psi = \frac{1}{r^2}\frac{\partial}{\partial r}r^2\frac{\partial\Psi}{\partial r} + \frac{1}{r^2\sin\theta}\frac{\partial}{\partial\theta}\sin\theta\frac{\partial\Psi}{\partial\theta} +$$

$$+\frac{1}{r^2\sin^2\theta}\frac{\partial^2\Psi}{\partial\phi^2}.$$

The components of the strain tensor are given by the set of equations

$$\varepsilon_{rr} = \frac{\partial u_r}{\partial r}$$

$$\varepsilon_{\theta\theta} = \frac{1}{2}\frac{\partial u_\theta}{\partial\theta} + \frac{u_r}{r}$$

$$\varepsilon_{\phi\phi} = \frac{1}{r\sin\theta}\cdot\frac{\partial u_\phi}{\partial\phi} + \frac{u_r}{r} + \frac{\cot\theta}{r}u_\theta$$

$$\varepsilon_{\theta\phi} = \frac{1}{2}\frac{1}{r}\frac{\partial u_\phi}{\partial\theta} - \frac{\cot\theta}{r}u_\phi + \frac{1}{r\sin\theta}\frac{\partial u_\theta}{\partial\phi}$$

$$\varepsilon_{\phi r} = \frac{1}{2}\left[\frac{1}{r\sin\theta}\cdot\frac{\partial u_r}{\partial\phi} - \frac{u_\phi}{r} + \frac{\partial u_\phi}{\partial r}\right]$$

$$\varepsilon_{\theta r} = \frac{1}{2}\left[\frac{1}{r}\frac{\partial u_r}{\partial\theta} - \frac{u_\theta}{r} + \frac{\partial u_\theta}{\partial r}\right]$$

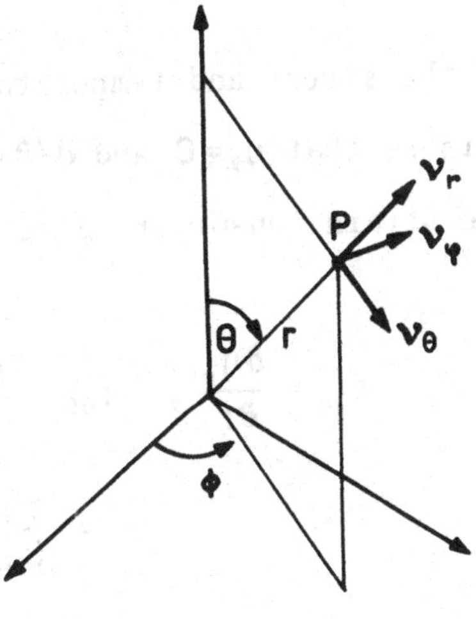

Fig. 2

and the equations of motion by the set

$$\frac{1}{r^2}\frac{\partial}{\partial r}(r^2\sigma_{rr}) + \frac{1}{r\sin\theta}\frac{\partial}{\partial\theta}(\sin\theta\,\sigma_{r\theta}) + \frac{\partial\sigma_{r\phi}}{\partial\phi} - \frac{\sigma_{\theta\theta}-\sigma_{\phi\phi}}{r} + \sigma X_r = \sigma\frac{\partial^2 u_r}{\partial t^2} \quad (1.8.14)$$

$$\frac{1}{r^2}\frac{\partial}{\partial r}(r^3\sigma_{r\theta})+\frac{1}{r\sin\theta}\frac{\partial}{\partial\theta}(\sin\theta\,\sigma_{\theta\theta})+\frac{\partial\sigma_{\theta\phi}}{\partial\phi}-\frac{\cot\theta}{r}\sigma_{\phi\phi}+\sigma X_\theta=\sigma\frac{\partial^2 u_\theta}{\partial t^2}$$

(1.8.15)

$$\frac{1}{r^2}\frac{\partial}{\partial r}(r^3\sigma_{r\phi})+\frac{1}{r\sin^2\theta}\frac{\partial}{\partial\theta}(\sin^2\theta\,\sigma_{\theta\phi})+\sigma X_\phi=\sigma\frac{\partial^2 u_\phi}{\partial t^2}\ . \qquad (1.8.16)$$

If the stress and temperature fields are symmetric about the z-axis so that $u_\phi = 0$ and $\partial/\partial\phi = 0$ the non-vanishing components of the strain tensor are given by the equations

$$\varepsilon_{rr}=\frac{\partial u_r}{\partial r}\ ,\qquad \varepsilon_{\theta\theta}=\frac{1}{r}\frac{\partial u}{\partial\theta}+\frac{u_r}{r}\ ,\qquad \varepsilon_{\phi\phi}=\frac{u_r}{r}+\cot\theta\,\frac{u_\theta}{r}=$$

$$=\frac{1}{2}\left[r\frac{\partial}{\partial r}\left(\frac{u_\theta}{r}\right)+\frac{1}{r}\frac{\partial u_r}{\partial\theta}\right]$$

so that the stress-strain relations become

$$\sigma_{rr}=\frac{E}{(1+\eta)(1-2\eta)}\left[(1-\eta)\frac{\partial u_r}{\partial r}+2\eta\frac{u_r}{r}+\frac{\eta}{r\sin\theta}\frac{\partial}{\partial\theta}(\sin\theta\,u_\theta)-\right.$$

$$\left.-(1+\eta)\alpha(T-T_0)\right] \qquad (1.8.17)$$

$$\sigma_{\theta\theta} = \frac{E}{(1+\eta)(1-2\eta)}\left[(1-\eta)\frac{1}{r}\frac{\partial u_\theta}{\partial\theta} + \frac{u_r}{r} + \eta\frac{\partial u_r}{\partial r} + \eta\frac{\cot\theta}{r}u_\theta - \right.$$

$$\left. - (1+\eta)\alpha(T-T_0)\right] \tag{1.8.18}$$

$$\sigma_{\phi\phi} = \frac{E}{(1+\eta)(1-2\eta)}\left[(1-\eta)\frac{\cot\theta}{r}u_\theta + \frac{u_r}{r} + \eta\frac{\partial u_r}{\partial r} + \frac{\eta}{r}\frac{\partial u_\theta}{\partial\theta} - \right.$$

$$\left. - (1+\eta)\alpha(T-T_0)\right] \tag{1.8.19}$$

$$\sigma_{r\theta} = \frac{E}{2(1+\eta)}\left[\frac{1}{r}\frac{\partial u_r}{\partial\theta} - \frac{u_0}{r} - \frac{\partial u_\theta}{\partial r}\right]. \tag{1.8.20}$$

The equations of motion reduce to

$$\frac{1}{r^2}\frac{\partial}{\partial r}r^2\sigma_{rr} + \frac{1}{r\sin\theta}\frac{\partial}{\partial\theta}(\sin\theta\,\sigma_{r\theta}) - \frac{\sigma_{\theta\theta}+\sigma_{rr}}{r} + \sigma X_r = \sigma\frac{\partial^2 u_r}{\partial t^2} \tag{1.8.21}$$

$$\frac{1}{r^3}\frac{\partial}{\partial r}r^3\sigma_{r\theta} + \frac{1}{r\sin\theta}\frac{\partial}{\partial\theta}(\sin\theta\,\sigma_{\theta\theta}) - \frac{\cot\theta}{r}\sigma_{\phi\phi} + \sigma X_\theta = \sigma\frac{\partial^2 u_\theta}{\partial t^2}\ . \tag{1.8.22}$$

In the case where there is spherical symmetry so that $u_\theta = u_\phi \equiv 0$ and all of the physical quantities depend only on

r and t the only non-vanishing components of the stress tensor are given by the equations

$$\sigma_{rr} = \frac{E}{(1+\eta)(1-2\eta)}\left[(1-\eta)\frac{\partial u_r}{\partial r} + 2\eta\frac{u_r}{r} - (1+\eta)\alpha(T-T_0)\right] \tag{1.8.23}$$

$$\sigma_{\theta\theta} = \frac{E}{(1+\eta)(1-2\eta)}\left[\eta\frac{\partial u_r}{\partial r} + \frac{u_r}{r} - (1+\eta)\alpha(T-T_0)\right] \tag{1.8.24}$$

$$\sigma_{\phi\phi} = \frac{E}{(1+\eta)(1-2\eta)}\left[\eta\frac{\partial u_r}{\partial r} + \frac{u_r}{r} - (1+\eta)\alpha(T-T_0)\right]. \tag{1.8.25}$$

If we substitute these expressions in the equation of motion

$$\frac{\partial \sigma_{rr}}{\partial r} + \frac{1}{r}(2\sigma_{rr} - \sigma_{\theta\theta} - \sigma_{\phi\phi}) + \sigma X_r = \sigma\frac{\partial^2 u_r}{\partial t^2} \tag{1.8.26}$$

we find that the displacement u_r satisfies

$$\frac{\partial^2 u_r}{\partial r^2} + \frac{2}{r}\frac{\partial u_r}{\partial r} - \frac{2u_r}{r^2} - \frac{1+\eta}{1-\eta}\alpha\frac{\partial T}{\partial r} = \frac{(1+\eta)(1-2\eta)}{(1-\eta)E}\sigma\left(\frac{\partial^2 u_r}{\partial t^2} - X_r\right). \tag{1.8.27}$$

CHAPTER 2

THERMOELASTIC WAVES

2.1 Units

We begin by writing the equation (1.5.9) in dimensionless form by introducing as unit of length

$$l^* = \frac{k}{\sigma c_\varepsilon V_T} \qquad (2.1.1)$$

and as unit of time

$$t^* = l^*/V_T . \qquad (2.1.2)$$

Then

$$\omega^* = \frac{1}{t^*} = \frac{\sigma c_\varepsilon V_T^2}{k} \qquad (2.1.3)$$

will be the characteristic frequency. We also introduce a *coupling constant* ε defined by the equation

$$\varepsilon = \frac{\alpha^2 T_0 V_T^2}{c_\varepsilon}\left(\frac{1+\eta}{1-\eta}\right)^2 = \frac{9\alpha^2 T_0}{\sigma^2 c_\varepsilon \varkappa^2 V_T^2} . \qquad (2.1.4)$$

Values of ω^* and ε for four metals are given in Table 1.

TABLE 1

Basic data for four metals (at 20 °C)

Quantity	Units	Aluminium	Copper	Iron	Lead
v_p	cm sec^{-1}	6.32×10^{5}	4.36×10^{5}	5.80×10^{5}	2.14×10^{5}
ε	—	3.56×10^{-2}	1.68×10^{-2}	2.97×10^{-2}	7.33×10^{-2}
ω^*	sec^{-1}	4.66×10^{11}	1.73×10^{11}	1.75×10^{11}	1.91×10^{11}
q_∞	cm^{-1}	1.31×10^{3}	3.29×10^{3}	4.48×10^{3}	3.27×10^{3}
ω_c	sec^{-1}	9.80×10^{13}	7.55×10^{13}	9.95×10^{13}	3.69×10^{13}

If we now write

$$\tau = t/t^*, \quad \Delta_3^* = l^{*2}\Delta_3$$

we find that equation (1.5.9) takes the dimensionless form

$$\left(\frac{\partial}{\partial\tau} - \Delta_3^*\right)\left(\frac{\partial^2}{\partial\tau^2} - \Delta_3^*\right)\phi = \varepsilon\frac{\partial}{\partial\tau}\Delta_3^*\phi\,. \tag{2.1.5}$$

2.2 Plane Harmonic Elastic Waves

We shall begin by considering the propagation of plane harmonic waves. A plane wave is characterized by the property that $\vec{u}$ and T depend only on the time and on the distance measured along a fixed propagation vector. We may, without loss of generality, take the x-axis along the direction and choose $u_x = u(x,t)$, $u_y = u_z = 0$. If the reference temperature T_0 is assumed to be uniform and if we write

$$\theta = T - T_0 \tag{2.2.1}$$

equations (1.5.3)and (1.5.4) reduce to the forms

$$\frac{\partial^2 u}{\partial t^2} = V_T^2 \frac{\partial^2 u}{\partial x^2} - \frac{\alpha}{3\sigma\varkappa}\frac{\partial \theta}{\partial x} \tag{2.2.2}$$

$$\sigma c_\varepsilon \frac{\partial \theta}{\partial t} - \frac{\alpha T_0}{3\varkappa}\frac{\partial^2 u}{\partial x \partial t} = k\frac{\partial^2 \theta}{\partial x^2} . \tag{2.2.3}$$

If we assume plane waves solutions

$$u = u_0 \exp\{i(px - \omega t)\} , \quad \theta = \theta_0 \exp\{i(px - \omega t)\} \tag{2.2.4}$$

we find that

$$(V_T^2 p^2 - \omega^2)u_0 + i(\alpha/\sigma\varkappa)p\theta_0 = 0$$

$$(\alpha T_0/\varkappa)p\omega u_0 + (kp^2 - i\sigma c_\varepsilon \omega)\theta_0 = 0 .$$

Writing

$$\omega = \omega^* \chi \qquad p = (\omega^*/V_T)\xi \tag{2.2.5}$$

where ω^* is defined by equation (2.1.3) we can transform these equations to

$$3\sigma\varkappa V_T \omega^* (\xi^2 - \chi^2)u_0 + i\alpha\xi\theta_0 = 0$$
$$\alpha T_0 V_T \xi \chi u_0 + 3k(\xi^2 - i\chi)\theta_0 = 0 . \tag{2.2.6}$$

Eliminating u_0 and θ_0 from these equations we derive the equation

$$(\xi^2 - \chi^2)(\chi + i\xi^2) + \varepsilon\chi\xi^2 = 0 , \tag{2.2.7}$$

where ε is the constant defined by equation (2.1.4), which expresses the relationship between ε and $\varkappa$.

We obtain "uncoupled" waves by taking $\alpha=0$, which is equivalent to taking $\varepsilon=0$; equation (2.2.7) then reduces to the simple form

$$(\xi^2-\varkappa^2)(\varkappa+i\xi^2)=0$$

from which we deduce immediately that it has the roots $\pm\xi_1^{(0)}$, $\pm\xi_2^{(0)}$ where

$$\xi_1^{(0)}=\varkappa\ ,\quad \xi_2^{(0)}=\frac{1}{2}(1+i)\sqrt{(2\varkappa)}\ . \tag{2.2.8}$$

If we take ξ to be a real constant we can regard the equation as being an equation in $\varkappa$ with roots $\pm\xi, -\xi^2$. Substituting these values into equations (2.2.4) and (2.2.5) we find that

$$u = u_0^+\exp\{i\eta(x-V_Tt)\}+u_0^-\exp\{i\eta(x+V_Tt)\} \tag{2.2.9}$$

$$\theta = \theta_0\exp\{-(k/\sigma c_\varepsilon)\eta^2t+i\eta x\}\ . \tag{2.2.10}$$

Equation (2.2.9) represents elastic waves moving along the x-axis, u_0^+ being the amplitude of the wave moving in the positive direction and u_0^- that of the wave moving in the negative direction. Equation (2.2.10) represents the thermal wave; it is a standing wave whose amplitude decays exponentially with the time.

On the other hand if we assume $\varkappa$, i.e. ω , to be a real constant we obtain the solutions

$$u = u_0^+ \exp\{i\omega(t - x/V_T)\} + u_0^- \exp\{i\omega(t + x/V_T)\}$$

$$\theta = \theta_0^+ \exp\{-\omega x/V + i\omega(t - x/V)\} + \theta_0^- \exp\{\omega x/V + i(t - x/V)\}$$

where the phase velocity, V , of the thermal waves are defined by the equation

$$V^2 = 2k\omega/(\varrho c_\varepsilon) .$$

It will be noted that the thermal wave is attenuated and, in the sense that its phase velocity is a function of the frequency, is subject to dispersion.

In the general case $\alpha \neq 0$, the roots of equation (2.2.7) are $\pm\xi_1, \pm\xi_2$ where

$$\begin{aligned} \xi_1 &= \frac{1}{2}\sqrt{\chi}\,\{\chi + (1+i)\sqrt{(2\chi)} + i(1+\varepsilon)\}^{\frac{1}{2}} + \{\chi - (1+i)\sqrt{(2\chi)} + i(1+\varepsilon)\}^{\frac{1}{2}} \\ \xi_2 &= \frac{1}{2}\sqrt{\chi}\,\{\chi + (1+i)\sqrt{(2\chi)} + i(1+\varepsilon)\}^{\frac{1}{2}} - \{\chi - (1+i)\sqrt{(2\chi)} + i(1+\varepsilon)\}^{\frac{1}{2}} . \end{aligned} \qquad (2.2.11)$$

Since $\xi_1 \to \xi_1^{(0)}$ as $\varepsilon \to 0$ we see that the roots $\pm\xi_1$ correspond to modified elastic waves; similarly, since $\xi_2 \to \xi_2^{(0)}$ as $\varepsilon \to 0$, the roots $\pm\xi_2$ correspond to modified thermal waves. Separating ξ_1 , ξ_2 into their real and imaginary parts, with the notation

$$\xi_j = V_T \frac{\chi}{v_j} + i\frac{q_j}{\omega^*} , \quad (j = 1,2) \qquad (2.2.12)$$

and introducing two functions θ_1 and u_1 through the pair of equations

$$\theta_1 = \frac{\alpha T_0 V_T^2}{k\varkappa} - \frac{\omega^* v_1 V_T(\omega^* \varkappa + i v_1 q_1)}{i\omega^{*2} v_1^2 \varkappa - V_T^2(\omega^* \varkappa + i v_1 q_1)^2}, \qquad (2.2.13)$$

$$u_1 = \frac{\alpha k}{\sigma^2 c_\varepsilon \varkappa V_T^2}, \frac{\omega^* v_2(\omega^* \varkappa + i v_2 q_2)}{V_T^2(\omega^* \varkappa + i v_2 q_2)^2 - \omega^* v_2^2 \varkappa^2}, \qquad (2.2.14)$$

we obtain the solution

$$\begin{aligned} u = {} & u_0^+ \exp\{-q_1 x - i\omega^* \varkappa(t - x/v_1)\} + u_0^- \exp\{q_1 x - i\omega^* \varkappa(t + x/v_1)\} \\ & - i u_1 [\theta_0^+ \exp\{-q_2 x - i\omega^* \varkappa(t - x/v_2)\} - \theta_0^- \exp\{q_2 x - i\omega^* \varkappa(t + x/v_2)\}], \end{aligned} \qquad (2.2.15)$$

$$\begin{aligned} \theta = {} & \theta_0^+ \exp\{-q_2 x - i\omega^* \varkappa(t - x/v_2)\} + \theta_0^- \exp\{q_2 x - i\omega^* \varkappa(t + x/v_2)\} \\ & + \theta_1 [u_0^+ \exp\{-q_1 x - i\omega^* \varkappa(t - x/v_1)\} - u_0^- \exp\{q_1 x - i\omega^* \varkappa(t + x/v_2)\}]. \end{aligned} \qquad (2.2.16)$$

In equations (2.2.15) and (2.2.16) we recognize two types of harmonic waves. Those with exponents $(t \pm x/v_1)$ are called *quasi-elastic waves* while those with exponents $(t \pm x/v_2)$ are called *quasi-thermal waves*.

The phase velocities v_1 , v_2 and the attenuation coefficients q_1 , q_2 can be calculated by means of equations (2.2.11) and (2.2.12).

If we expand in powers of ε we obtain the formulae

$$\nu_1 = V_T\left[1+\frac{1}{2(1+x^2)}\varepsilon-\frac{1-14x^2+x^4}{8(1+x^2)^2}\varepsilon^2+\right. \tag{2.2.17}$$

$$\left.+\frac{1-79x^2+159x^4-17x^6}{16(1+x^2)^5}\varepsilon^3+0(\varepsilon^4)\right],$$

$$\nu_2 = \sqrt{(2x)}\,V_T\left[1-\frac{1+x}{2(1+x^2)}\varepsilon+\frac{3+10x-8x^2-6x^3+5x^4}{8(1+x^2)^3}\varepsilon^2-\right.$$

$$\left.-\frac{5+35x-65x^2-151x^3+143x^4+73x^5-43x^6+3x^7}{16(1+x^2)^5}+0(\varepsilon^4)\right], \tag{2.2.18}$$

$$q_1 = \frac{\omega^*}{V_T}\left[\frac{x^2}{2(1+x^2)}\varepsilon-\frac{x^2(5-3x^2)}{4(1+x^2)^3}\varepsilon^2+\right. \tag{2.2.19}$$

$$\left.+\frac{x^2(35-155x^2+65x^4-x^6)}{16(1+x^2)^5}\varepsilon^3+0(\varepsilon^4)\right],$$

$$q_2 = \frac{\omega^*}{V_T}\sqrt{\left(\frac{1}{2}x\right)}\left[1+\frac{1-x}{2(1+x^2)}\varepsilon-\frac{1-6x-12x^2+10x^3+3x^4}{8(1+x^2)^3}\varepsilon^2+\right. \tag{2.2.20}$$

$$\left.+\frac{1-15x-65x^2+135x^3+155x^4-101x^5-35x^6+5x^7}{16(1+x^2)^5}\varepsilon^3+0(\varepsilon^4)\right].$$

From equation (2.2.18) we see that q_1, the attenuation coefficient of the quasi-elastic mode, varies like $\varkappa^2$ at low frequencies and, as $\varkappa$ tends to infinity, tends to the limiting value

$$q_\infty = \frac{1}{2}\varepsilon\frac{\omega^*}{V_T}. \tag{2.2.21}$$

It should be noted that if $\varkappa = 1$,

$$q_1 = \frac{1}{2}q_\infty\left[1+O(\varepsilon)\right] \tag{2.2.22}$$

values of q_∞ for four metals are included in Table 1 above.

Also included in Table 1 are the relevant values of the frequency

$$\omega_c = 2\pi V_P\left(\frac{3\sigma}{4\pi M}\right)^{\frac{1}{3}} \tag{2.2.23}$$

of the Debye spectrum, which is the upper limit of the range of frequencies which can be attained in an elastic solid; here V_P denotes the velocity of longitudinal elastic waves and M is the mass of a constituent atom of the solid. From these data it will be seen that the quasi-elastic mode is very severely attenuated at frequencies in the neighbourhood of the characteristic value ω^*, and that $\omega^*/\omega_c < 0.01$. This implies that the upper limit of attainable frequencies in a solid is imposed by the thermoelastic damping and not by the atomic structure. Because of the presence of impurities and other scattering centres, the highest frequencies attainable in a coherent pulse are in practice very much smaller than ω^*. In practical applications of the theory we therefore have

$$\varkappa \ll 1 . \tag{2.2.24}$$

For that reason it is often more convenient to use power series in $\varkappa$ rather than ε to express approximate results for the phase velocities and the attenuation constants. These are easily shown to be

$$v_1 = V_P\left[1-\frac{\varepsilon(4-3\varepsilon)}{8(1+\varepsilon)^4}\varkappa^2+\frac{\varepsilon(64-304\varepsilon+232\varepsilon^2-17\varepsilon^3)}{128(1+\varepsilon)^8}\varkappa^4+0(\varkappa^6)\right] \tag{2.2.25}$$

$$v_2 = V_P(1+\varepsilon)^{-1}\sqrt{(2\varkappa)}\left[1-\frac{\varepsilon}{2(1+\varepsilon)^2}\varkappa+\frac{\varepsilon(4+\varepsilon)}{8(1+\varepsilon)^4}\varkappa^2+\right.$$
$$\left.+\frac{(8-20\varepsilon+\varepsilon^2)}{16(1+\varepsilon)^6}\varkappa^3+0(\varkappa^4)\right] \tag{2.2.26}$$

where

$$V_P = V_T\sqrt{(1+\varepsilon)} = \left[\frac{(1+\varepsilon)(1-\eta)}{(1+\eta)(1-2\eta)}\,\frac{E}{\sigma}\right]^{\frac{1}{2}} \tag{2.2.27}$$

is the velocity of longitudinal elastic waves (*), and

$$q_1 = \frac{\omega^*}{V_P}\left[\frac{\varepsilon}{2(1+\varepsilon)^2}\varkappa^2-\frac{\varepsilon(8-20\varepsilon+5\varepsilon^2)}{16(1+\varepsilon)^6}\varkappa^4+0(\varkappa^6)\right] \tag{2.2.28}$$

(*) It will be recalled that E and η are the *isothermal* elastic constants.

$$q_2 = \frac{\omega^*}{V_P}(1+\varepsilon)\sqrt{\left(\frac{1}{2}\right)}\left[1-\frac{\varepsilon}{2(1+\varepsilon)^2}\chi-\frac{\varepsilon(4-\varepsilon)}{8(1+\varepsilon)^4}\chi^2+\frac{\varepsilon(8-12\varepsilon+\varepsilon^2)}{16(1+\varepsilon)^6}\chi^3+0(\chi^4)\right]. \tag{2.2.29}$$

We notice that, in particular, equation (2.2.25)

$$\nu_1 \sim V_P \qquad \chi \longrightarrow 0$$

while equation (2.2.17)

$$\nu_1 \sim V_T \qquad \chi \longrightarrow \infty .$$

Values of q_1 over the frequency range $10^{-11} < \chi < 10^3$, for each of the metals listed in Table 1, have been calculated by Chadwick and Sneddon (1958) from equations (2.2.11) and (2.2.12. The results for copper are summarized in Table 2 below.

TABLE 2

Variation with frequency of the attenuation coefficient and phase velocity of the quasi-elastic mode in copper at 20 °C

χ	q_1/q_∞	ν_1/ν_T	χ	q_1/q_∞	ν_1/ν_T
10^{-2}	0.0001	1 0084	3	0.9033	1.0009
10^{-1}	0.0095	1 0083	7	0.9822	1.0002
0.3	0.0800	1 0077	10	0.9921	1.0001
0.7	0.3250	1 0057	10^2	1.0000	1.0000
1	0.4987	1 0043			

The variation of q_1/q_∞ with χ is plotted in Fig. 1 on a log-log scale. From this diagram we see that q_1 increases rapidly with χ

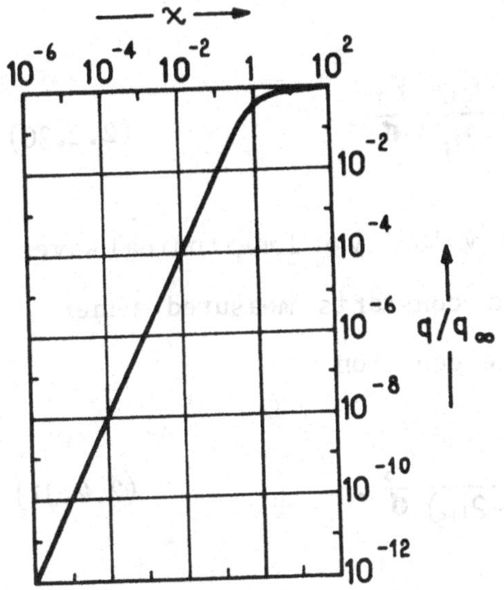

Fig. 1

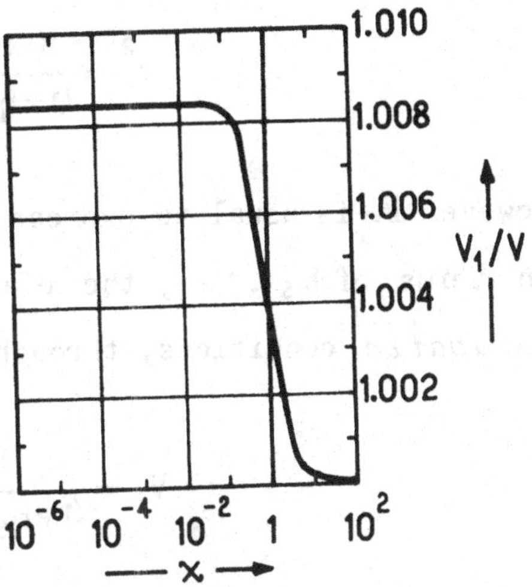

Fig. 2

until in the neighbourhood of ω^*, the gradient falls off sharply and the curve levels out.

The variation of the phase velocity v_1 with $\varkappa$ may also be calculated by means of equations (2.2.11) and (2.2.12). The numerical results obtained by Chadwick and Sneddon are included in Table 2, and the variation of v/V_T with $\varkappa$ shown (on a log-linear scale) in Fig. 2.

From equation (2.2.25) we see that at the frequencies normally occurring either naturally or under laboratory conditions (i.e. obeying the condition (2.2.24) $v_1 = V_P$. The velocity V_p is called the velocity of *longitudinal* elastic waves. We recall that in the above analysis we dropped, for convenience, the suffix from both E and n ; if we restore it then we may rewrite equation (2.27) in the form

$$V_P^2 = \frac{(1+\varepsilon)(1-\eta_T)}{(1+\eta_T)(1-2\eta_T)} \frac{E_T}{\sigma} . \tag{2.2.30}$$

However it is usual to express the velocity of longitudinal waves in terms of E_S , η_S , the elastic constants measured under *adiabatic* conditions, through the equation

$$V_P^2 = \frac{(1-\eta_S)}{(1+\eta_S)(1-2\eta_S)} \frac{E_S}{\sigma} . \tag{2.2.31}$$

Similarly, from equation (1.5.8) we see that V_S the velocity of *transverse* waves is defined in terms of the i-sothermal elastic constants by the equation

$$V_S^2 = \frac{E_T}{2\sigma(1+\eta_T)} . \tag{2.2.32}$$

If we define it in terms of the adiabatic constants by

$$V_S^2 = \frac{E_S}{2\sigma(1+\eta_S)} \tag{2.2.33}$$

it follows immediately that

$$\frac{E_S}{1+\eta_S} = \frac{E_T}{1+\eta_T} .$$

Equating the right hand sides of equations (2.2.30) and (2.2.31) we obtain the relation

$$\frac{1-\eta_T}{1-2\eta_T}(1+\varepsilon) = \frac{1-\eta_S}{1-2\eta_S}$$

which may be solved for η_S to give

$$\eta_S = \eta_T + \frac{\varepsilon(1-\eta_T)(1-2\eta_T)}{1+2\varepsilon(1-\eta_T)} \quad . \qquad (2.2.34)$$

From this it follows that

$$\frac{E_S}{E_T} = 1 + \frac{\varepsilon(1-\eta_T)(1-2\eta_T)}{(1+\eta_T)\{1+2\varepsilon(1-\eta_T)\}} \quad . \qquad (2.2.35)$$

We now turn our attention to the *quasi-thermal mode*, but before doing so look at the form of waves of assigned length. If we take ξ to be a real constant, equation (2.2.7) is a cubic equation in (2.2.7) whose three roots may be written as

$$x_1 = f - ig \,, \quad x_2 = -f - ig \,, \quad x_3 = -ih \,, \qquad (2.2.36)$$

where g is the *real* root of the cubic

$$8g^3 - 8g^2\xi^2 + 2g\xi^2(\xi^2+1+\varepsilon) - \varepsilon\xi^4 = 0 \qquad (2.2.37)$$

and

$$h = \xi^2 - 2g \,, \quad f^2 = 3g - 2\xi^2 g + \xi^2(1+\varepsilon) \,. \qquad (2.2.38)$$

In the limiting case $\varepsilon = 0$, the only real root of (2.2.37) is

$g = 0$ and we have $\varkappa_1 = \xi$, $\varkappa_2 = -\xi$, $\varkappa_3 = -i\xi^2$. It follows from the considerations that led to equations (2.2.9) and (2.2.10) that we may identify the roots $\varkappa_1, \varkappa_2$ with the quasi-elastic mode and $\varkappa_3$ with the quasi-thermal mode.

From equations (2.2.4) and (2.2.36) we deduce the following plane wave solution of the thermoelastic equations:

$$u = u_0^+ \exp\{-\omega^* gt + i(\omega^* \xi / V_T)(x - fV_T t/\xi)\}$$

$$+ u_0^- \exp\{-\omega^* gt + i(\omega^* \xi / V_T)(x + fV_T t/\xi)\}$$

$$- i\theta_0 \frac{\alpha k}{\sigma^2 c_\varepsilon \varkappa V_T^3} \cdot \frac{\xi}{\xi^2 + h^2} \exp\{-\omega^* ht + i\omega^* \xi x / V_T\} \tag{2.2.39}$$

$$\theta = -\frac{i\alpha T_0 V_T}{k\varkappa}\left[\frac{\xi(f - ig)}{f - i(g - \xi^2)} u_0^+ \exp\{-\omega^* gt + i(\omega^* \xi/V_T)(x - fV_T t/\xi)\}\right.$$

$$\left. + \frac{\xi(f + ig)}{f + i(g - \xi^2)} u_0^- \exp\{-\omega^* gt + i(\omega^* \xi/V_T)(x + fV_T t/\xi)\}\right]$$

$$+ \theta_0 \exp\{-\omega^* ht + i\omega^* \xi x / V_T\} . \tag{2.2.40}$$

These equations show the same properties of modification and coupling observed above for thermo-elastic waves of constant frequency. Like the purely thermal distance, the quasi-thermal mode is a standing wave.

As in the case of waves of constant frequency we

an derive power series in ε :

$$= \xi\left[1+\frac{1}{2(1+\xi^2)}\varepsilon-\frac{1-6\xi^2+\xi^4}{8(1+\xi^2)^3}\varepsilon^2+\frac{1-25\xi^2+35\xi^4-3\xi^6}{16(1+\xi^2)^5}\varepsilon^3+O(\varepsilon^4)\right], \quad (2.2.41)$$

$$g=\frac{\xi^2}{2(1+\xi^2)}\varepsilon-\frac{\xi^2(1-\xi^2)}{2(1+\xi^2)^3}\varepsilon^2+\frac{\xi^2(1-5\xi^2+2\xi^4)}{2(1+\xi^2)^5}\varepsilon^3+O(\varepsilon^4), \quad (2.2.42)$$

$$h=\xi^2\left[1-\frac{1}{1+\xi^2}\varepsilon-\frac{1-\xi^2}{(1+\xi^2)^3}\varepsilon^2-\frac{1-5\xi^2+\xi^4}{(1+\xi^2)^5}\varepsilon^3+O(\varepsilon^4)\right]. \quad (2.2.43)$$

ach of these series converges uniformly in ε for all positive alues of ξ and the boundedness of the coefficients makes them uitable for computation. However, in many particular problems f practical interest in which ξ is small it is preferable to se the power series expansions in ξ. These are

$$f=\xi\sqrt{(1+\varepsilon)}\left[1-\frac{\varepsilon(4+\varepsilon)}{8(1+\varepsilon)^3}\xi^2+\frac{\varepsilon(64-48\varepsilon-8\varepsilon^2-\varepsilon^3)}{128(1+\varepsilon)^6}\xi^4+O(\xi^6)\right] \quad (2.2.44)$$

$$g=\frac{\xi^2\varepsilon}{2(1+\varepsilon)}\left[1-\frac{1}{(1+\varepsilon)^3}\xi^2+\frac{1-2\varepsilon}{(1+\varepsilon)^6}\xi^4+O(\xi^6)\right] \quad (2.2.45)$$

$$h=\frac{\xi^2}{1+\varepsilon}\left[1+\frac{\varepsilon}{(1+\varepsilon)^3}\xi^2-\frac{(1-2\varepsilon)}{(1+\varepsilon)^6}\xi^4+O(\xi^6)\right]. \quad (2.2.46)$$

In equations (2.2.39) and (2.2.40) the amplitude of the quasi-thermal standing wave is proportional to $\exp(-t/\tau)$ where τ , the decay time, is given by the formula

$$\tau = \frac{1}{h\omega^*} .$$

If we denote the wave-length by λ then $(\omega^* \xi \lambda / V_T) = 2\pi$ and introduce a constant β defined by

$$\beta = \frac{\xi^2}{h} - 1$$

we find that

$$\tau = \frac{\sigma c_\varepsilon (1+\beta)\lambda^2}{4\pi^2 k} . \qquad (2.2.47)$$

From equation (2.2.46) we deduce that

$$\beta = \varepsilon \left[1 - \frac{1}{(1+\varepsilon)^2} + \frac{\varepsilon(1-\varepsilon)}{(1+\varepsilon)^5} \xi^4 - \frac{\varepsilon(1-4\varepsilon+2\varepsilon^2)}{(1+\varepsilon)^8} \xi^6 + O(\xi^8) \right] \qquad (2.2.48)$$

which is a suitable expansion for β when ξ is very small. When ξ is not small we may make use of the formula

$$\beta = \frac{\varepsilon}{1+\xi^2} \left[1 + \frac{2\xi^2}{(1+\xi^2)^2} - \frac{(3-5\xi^2)\xi^2}{(1+\xi^2)^4} \varepsilon^2 + O(\varepsilon^3) \right] \qquad (2.2.49)$$

which is easily derived from equation (2.2.43). From this we deduce immediately that

$$\beta \sim \varepsilon \xi^{-2} \qquad \xi \longrightarrow \infty .$$

The value $\xi = 1$ corresponds to wavelength of order 10^{-5} cm. and decay times of the order of 10^{-12} sec. From this we see that temperature disturbances of practical interest will satisfy the condition $\xi \ll 1$. When this condition is satisfied, equation (2.2.48) gives $\beta \simeq \varepsilon$ and equation (2.2.47) reduces to

$$\tau \simeq \frac{\sigma_\varepsilon (1+\varepsilon)\lambda^2}{4\pi^2} .$$

The variation with wave number of the decay parameter β of the quasi-thermal mode in copper at 20° C.is shown in Table 3 and, graphically in Fig. 3. This curve is very similar in form to Fig. 2, the rapid change in slope of the curve occurring in the vicinity of $\xi = 1$.

TABLE 3

Variation with wave number of the decay parameter β of the quasi--thermal mode in copper at 20 °C

ξ	β	ξ	β
10^{-2}	0.01680	3	0.00169
10^{-1}	0.01664	7	0.00034
0.3	0.01580	10	0.00017
0.7	0.01136	10^2	0.00000
1	0.00847		

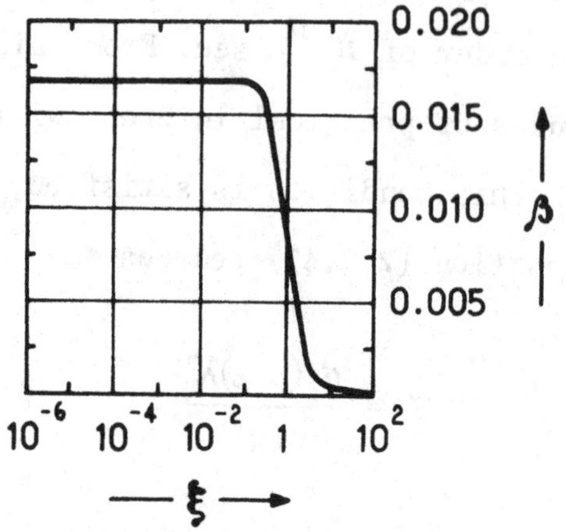

Fig. 3

2.3 Coupling

If we take $u_0^- = 0$, $\theta_0 = 0$ in equations (2.2.39) and (2.2.40) and write u_0^+ simply as u_0 we obtain the simple solution

$$u = u_0 \exp\{-\omega^* gt + i(\omega^* \xi / V_T)(x - fV_T t/\xi)\} \qquad (2.3.1)$$

$$\theta = -\frac{i\alpha T_0 V_T}{k\varkappa} \cdot \frac{\xi(f-ig)}{f-i(g-\xi^2)} u_0 \exp\{-\omega^* gt + i(\omega^* \xi/V_T)(x - fV_T t/\xi)\} .$$

If we substitute

$$-\frac{\xi(f-ig)}{f-i(g-\xi^2)} = Ae^{i(\pi+\gamma)}$$

where

$$A = \xi\left[\frac{f^2+g^2}{f^2+(g-\xi^2)^2}\right]^{\frac{1}{2}}, \qquad \tan\gamma = \frac{f^2+g^2-g\xi^2}{f\xi^2} \qquad (2.3.2)$$

in this last equation for θ and equate real parts we find that

$$\theta = \frac{\alpha T_0 V_T}{k\varkappa} A \exp(-\omega^* gt)\cos\{(\omega^*\xi/V_T)(x-fV_Tt/\xi)+\pi+\gamma\}. \qquad (2.3.3)$$

The pair of equations (2.3.1), (2.3.3) represent a thermoelastic wave in which the quasi-thermal mode is absent travelling along the x-axis in the positive direction. The factor A describes the coupling between the elastic wave and the accompanying thermal disturbance and $\pi+\gamma$ is the difference in phase between the thermal and the elastic components.

Table 4 shows the variation of A and γ with reduced wave number in the range $10^{-2} < \xi < 10^2$. The variation of these parameters with ranging over a more extended set of values

TABLE 4

Variation with wave number of the coupling factor A and the phase angle γ of the quasi-elastic mode in copper at 20 °C

ξ	A	γ(degrees)	ξ	A	γ(degrees)
10^{-2}	0.0100	89.43	3	0.9501	18.32
10^{-1}	0.0995	84.34	7	0.9903	8.07
0.3	0.2877	73.41	10	0.9952	5.66
0.7	0.5756	55.09	10^2	1.0000	0.57
1	0.7101	45.00			

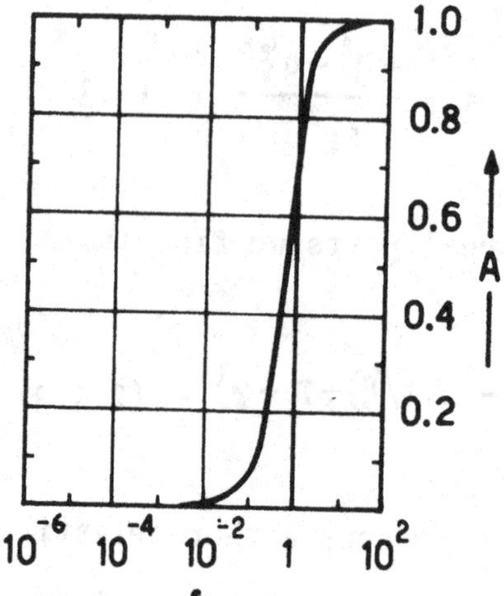

Fig. 4

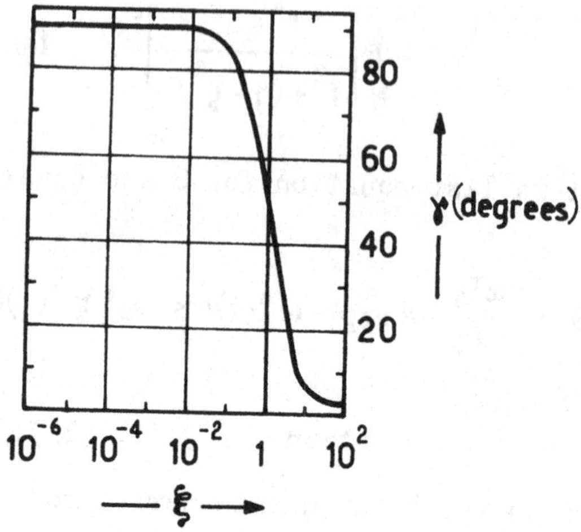

Fig. 5

is shown in Figs. 4 and 5. Again in these curves we get the now familiar sudden change in behaviour in the vicinity of $\xi = 1$.

To obtain the asymptotic values of A and γ we use the formulae

$$A = \xi(1+\xi^2)^{\frac{1}{2}}\left[1+\frac{\xi^2}{(1+\xi^2)^2}\varepsilon-\frac{\xi^2(3-4\xi^2)}{2(1+\xi^2)^4}\varepsilon^2+\frac{\xi^2(4-19\xi^2+10\xi^4)}{2(1+\xi^2)^6}\varepsilon^3+O(\varepsilon^4)\right] \tag{2.3.4}$$

$$\cot\gamma = \xi\left[1-\frac{1-\xi^2}{2(1+\xi^2)}\varepsilon+\frac{3-16\xi^2-\xi^4+2\xi^6}{8(1+\xi^2)^3}\varepsilon^2+O(\varepsilon^3)\right] \tag{2.3.5}$$

which can be derived immediately from equations (2.2.41) and

(2.2.42).

From these equations we see that

$$A \sim 1, \quad \gamma \sim 0 \quad \text{as} \quad \xi \to \infty .$$

Similarly, from equations (2.2.44) and (2.2.45) and the series

$$\gamma = \frac{1}{2}\pi - \cot\gamma + \frac{1}{3}\cot^3\gamma - \frac{1}{5}\cot^5\gamma + \ldots$$

which is valid for small values of $\cot\gamma$ we deduce the pair of equations

$$A = \xi\left[1 - \frac{1}{2(1+\varepsilon)^2}\xi^2 + \frac{3-5\varepsilon}{8(1+\varepsilon)^5}\xi^4 + 0(\xi^6)\right] \qquad (2.3.6)$$

$$\gamma = \frac{1}{2}\pi - (1+\varepsilon)^{\frac{1}{2}}\xi 1 - \frac{8-8\varepsilon-\varepsilon^2}{24(1+\varepsilon)^3}\xi^2 + \frac{128-768\varepsilon+288\varepsilon^2+32\varepsilon^3+3\varepsilon^4}{640(1+\varepsilon)^6}\xi^4 + 0(\xi^6) \qquad (2.3.7)$$

which can be used in the practical case when $\xi \ll 1$. From this pair of equations we see that

$$A \sim \xi \qquad \gamma \sim \frac{1}{2}\pi - (1+\varepsilon)^{-\frac{1}{2}}\xi \quad \text{as} \quad \xi \to 0 .$$

From the numerical calculations we see that *increases monotonically* from 0 to 1 and $\pi + \gamma$ *decreases monotonically* from $3\pi/2$ to π as ξ increases from 0 to ∞.

We see from Table 1 that for the metals listed

there $\alpha T_0 V_T/k\varkappa$ is of order 10^9(°C/cm) at 20°C. Thus as $\xi \to 1$ the coupling between the elastic and thermal components of the thermoelastic wave is extremely strong. On the other hand if $\xi << 1$ it follows from the above considerations that, if θ is measured in °C, $|\theta|$ is of the order of $\xi u_0 10^9$. Now from equation (2.3.1)

$$\left|\frac{\partial u}{\partial x}\right| = \frac{\omega^* \xi}{V_T} u_0$$

so that $|\theta|$ is of the order of

$$10^9 \frac{V_T}{\omega^*} \left|\frac{\partial u}{\partial x}\right| .$$

Since V_T/ω^* is of the order of 10^{-6} we see that $|\theta|$ is of the order of 10^3 times the maximum strain and therefore will usually amount to no more than a fraction of a degree. Under normal laboratory conditions, therefore, the coupling between a compressional elastic wave and the associated thermal wave will be small and the two waves will be out of phase by approximately $3\pi/2$. Although the coupling is small it should be well within the range of actual measurement.

2.4 Thermoelastic Rayleigh Waves

We now consider surface waves propagated in the x-direction along the surface of the half-space $z > 0$ which is occupied by elastic material which is free to exchange heat by radiation into an atmosphere $z<0$ maintained at a temperature T_0 (Lockett, 1958). If, therefore we write $\Theta = T - T_0$ we have the radiation condition

$$\frac{\partial \Theta}{\partial z} + h\Theta = 0 \quad \text{on} \quad z = 0 . \tag{2.4.1}$$

We shall also assume that before the propagation of the thermoelastic disturbance the temperature of the elastic solid is T_0 throughout and that during the disturbance the boundary is free from stress, i.e. that

$$\sigma_{xz} = \sigma_{yz} = \sigma_{zz} \quad \text{on} \quad z = 0 . \tag{2.4.2}$$

If we assume a displacement field of the form

$$u_x = \frac{\partial \phi}{\partial x} - \frac{\partial \Psi}{\partial z}, \quad u_y = 0, \quad u_z = \frac{\partial \phi}{\partial z} + \frac{\partial \Psi}{\partial x}, \tag{2.4.3}$$

where ϕ and Ψ are functions of x and z only, it follows from equations (1.5.3) and (1.5.4) that ϕ , Ψ and Θ satisfy the set of equations

$$\frac{\partial^2 \phi}{\partial t^2} = V_T^2 \Delta_2 \phi - \frac{\alpha}{\sigma \varkappa} \Theta \tag{2.4.4}$$

$$\sigma c_\varepsilon \frac{\partial \Theta}{\partial t} + \frac{\alpha T_0}{\varkappa} \frac{\partial}{\partial t} \Delta_2 \phi = k \Delta_2 \Theta \tag{2.4.5}$$

$$\frac{\partial^2 \Psi}{\partial t^2} = V_S^2 \Delta_2 \Psi \tag{2.4.6}$$

where in this instance

$$\Delta_2 = \frac{\partial^2}{\partial x^2} - \frac{\partial^2}{\partial z^2} .$$

We try to find plane wave solutions of these equations of the forms

$$\begin{aligned} \theta &= \Theta(z)\exp\{i(px-\omega t)\} \\ \phi &= \Phi(z)\exp\{i(px-\omega t)\} \\ \psi &= \Psi(z)\exp\{i(px-\omega t)\} . \end{aligned} \tag{2.4.7}$$

Substituting these expressions into equations (2.4.4)-(2.4.6) we see that the functions Θ, Φ and Ψ must satisfy the system of second order ordinary differential equations

$$\begin{aligned} &\Phi''(z) - (p^2 - \omega^2/V_T^2)\Phi = (\alpha/\sigma \varkappa V_T^2)\Theta \\ &\Theta''(z) - (p^2 - i\omega\sigma c_\varepsilon/k) = -i(\omega\alpha T_0/\varkappa k)(\Phi'' - p^2\Phi) \\ &\Psi''(z) - (p^2 - \omega/V_S^2)\Psi = 0 . \end{aligned}$$

In addition to the conditions (2.4.1) and (2.4.2) we have the additional condition that the solutions will tend to zero as $z \to \infty$. We therefore take solutions of equations (2.4.8) of the form

$$\Theta(z) = \frac{\sigma\varkappa V_T^2}{\alpha}\left[A(\omega^2/V_T^2 - \zeta_1^2)\exp\{-z\sqrt{(p^2-\zeta_1^2)}\} + \right.$$

$$+ B(\omega^2/V_T^2 - \zeta_2^2)\exp\{-z\sqrt{(p^2-\zeta_2^2)}\}]$$

$$\Phi(z) = A\exp\{-z\sqrt{(p^2-\zeta_1^2)}\} + B\exp\{-z\sqrt{(p^2-\zeta_2^2)}\}$$

$$\Psi(z) = C\exp\{-z\sqrt{(p^2-\zeta_3^2)}\}$$

where ζ_1^2, ζ_2^2 are the roots of the equation

$$\zeta^4 - \{\omega^2/V_T^2 + i(\omega\sigma c_\varepsilon/k)(1+\varepsilon)\}\zeta^2 + i\omega^3\sigma c_\varepsilon/kV_T^2 = 0 , \qquad (2.4.8)$$

and

$$\zeta_3^2 = \omega^2/V_S^2 \qquad (2.4.9)$$

and it is assumed that whenever $\sqrt{(p^2-\zeta^2)}$ occurs the root with positive real part is taken.

If we substitute from equations (2.4.3) into equations (1.4.4) we find that

$$\sigma_{zz} = \sigma V_T^2 \Delta_2\phi + 2\sigma V_S^2\left(\frac{\partial^2\Psi}{\partial x\partial z} - \frac{\partial^2\phi}{\partial z^2}\right) - \frac{\alpha\Theta}{3\varkappa}$$

$$\sigma_{xz} = \sigma V_S^2\left(2\frac{\partial^2\phi}{\partial x\partial y} + \frac{\partial^2\Psi}{\partial x^2} - \frac{\partial^2\Psi}{\partial z^2}\right)$$

so that the boundary conditions (2.4.2) reduce respectively to

$$\sigma V_T^2\{\Phi''(0) - p^2\Phi(0)\} - 2\sigma V_S^2\{\Phi''(0) + ip\Psi'(0)\} - (\alpha/3\varkappa)\Theta = 0$$

$$\{\Psi''(0)+p^2\Psi(0)\}-2ip\Phi'(0) = 0 .$$

The radiation condition becomes

$$\Theta'(0)+h\Theta(0) = 0 .$$

The constants A, B, C occurring in the above solution must therefore be connected through the equations

$$(A+B)(2-\omega^2/V_S^2p^2)-2i\beta_3 C = 0 ,$$

$$2i(\beta_1 A+\beta_2 B)+(2-\omega^2/V_S^2p^2)C = 0 ,$$

$$A(h/p-\beta_1)(\beta_1^2-1+\omega^2/V_T^2p^2)+B(h/p-\beta_2)(\beta_2^2-1+\omega^2/V_T^2p^2) = 0$$

where the β_i are defined by the equations

$$\beta_i^2 = 1-\zeta_i^2/p^2 , \quad (i=1,2,3) \tag{2.4.10}$$

Eliminating A, B, C from among these equations we obtain the relation

$$(2-\omega^2/V_S^2p^2)^2(\beta_1^2+\beta_1\beta_2+\beta_2^2-1+\omega^2/V_T^2p^2)-4\beta_1\beta_2\beta_3(\beta_1+\beta_2) =$$

$$= (h/p)\{(2-\omega^2/V_S^2p^2)(\beta_1+\beta_2)-4\beta_3(\beta_1\beta_2+1-\omega^2/V_T^2p^2)\} \tag{2.4.11}$$

connecting the wave number p and the frequency of the wave ω. If we take ω to be the independent variable in equations (2.4.7) and write

$$\nu = \frac{\omega}{p} \tag{2.4.12}$$

this equation becomes

$$(2-\nu^2/V_S^2)(\beta_1^2+\beta_1\beta_2+\beta_2^2-1+\nu^2/V_T^2)-4\beta_1\beta_2\beta_3(\beta_1+\beta_2) =$$

$$= (h\nu/\omega)\left[(\beta^1+\beta^2)(2-\nu^2/V_S^2)-4\beta_3(\beta_1\beta_2+1-\nu^2/V_T^2)\right] . \tag{2.4.13}$$

Now from the definitions (2.4.10)

$$\beta_1^2+\beta_2^2 = 2-(\zeta_1^2+\zeta_2^2)/p^2 \qquad \beta_1^2\beta_2^2 = 1-(\zeta_1^2+\zeta_2^2)/p^2+\zeta_1^2\zeta_2^2/p^4$$

and from equation (2.4.8)

$$\zeta_1^2+\zeta_2^2 = \omega^2/V_T^2+i(\omega\sigma c_\varepsilon/k)(1+\varepsilon) , \qquad \zeta_1^2\zeta_2^2 = i\omega^3\sigma c_\varepsilon/kV_T^2 .$$

Now by equations (2.1.3), (2.2.5) and (2.4.12)

$$\frac{\omega\sigma c_\varepsilon}{kp^2} = \frac{1}{\chi}\frac{\nu^2}{V_T^2} , \qquad \frac{\omega^3\sigma c_\varepsilon}{kp^4 V_T^2} = \frac{1}{\chi}\frac{\nu^4}{V_T^4}$$

so that we have the simple relations

$$\beta_1^2+\beta_2^2 = 2-\left(\frac{\nu}{V_T}\right)^2-\frac{i}{\chi}\left(\frac{\nu}{V_T}\right)^2(1+\varepsilon) \tag{2.4.14}$$

$$\beta_1^2\beta_2^2 = 1-\left(\frac{\nu}{V_T}\right)^2-\frac{i}{\chi}\left(\frac{\nu}{V_T}\right)^2(1+\varepsilon)+\frac{i}{\chi}\left(\frac{\nu}{V_T}\right)^4 \tag{2.4.15}$$

from which, with the addition of $\beta_3^2 = 1-(\nu/V_S)^2$, we may rid equation (2.4.13) of β's and obtain an equation for ν. From the original forms (2.4.7) we see that $1/\mathrm{Re}\,\nu^{-1}$ is the phase velocity and $\omega \mathrm{Im}\,\nu^{-1}$ is a measure of the attenuation in the x-direction. The thermoelastic surface wave is therefore subject to dispersion since both of these quantities depend on the frequency ω.

If we expand in powers of $\varkappa$ and neglect terms of order $\varkappa^{\frac{1}{2}}$ we find that the velocity ν ceases to depend upon the frequency and the thermal constant h ; we then have

$$(2-\nu^2/V_S^2)^2 = 4(1-\nu^2/V_S^2)\{1-\nu^2/(1+\varepsilon)V_T^2\}^{\frac{1}{2}}$$

which, if we write $V_P^2 = (1+\varepsilon)V_T^2$ is the well-known relation connecting ν, the velocity of Rayleigh waves, with the velocities of longitudinal and transverse elastic waves.

2.5 The Propagation of Thermal Stresses in Thin Metallic Rods

We now look at what is probably the simplest physicalsystem in which thermoelastic phenomena can occur - a thin metallic rod; in this case the basic equations assume their simplest forms. The solutions of problems relating to even this simple system are of physical interest and the methods used to obtain these solutions might be expected to have obvious generalizations in more complicated situations. We follow Sneddon (1958).

If we measure position along the rod by means of a coordinate x and denote the displacement of a point with coordinate x by u then the total strain at this point is given by $\varepsilon = \partial u/\partial x$. This total strain is made up of the elastic strain σ/E and the thermal strain $\alpha\theta$, where σ denotes the stress at the point x and $\theta = T - T_0$.

It is convenient to write the basic thermoelastic equations in dimensionless form. If we take a typical length l as our unit of length, a time τ as our unit of time, the reference temperature T as our unit of time, the reference temperature T as our unit of temperature, and the Young's modulus E as unit of stress, we find that equations (1.3.4), (1.4.4) and (1.5.4) respectively assume the forms

$$\frac{\partial \sigma}{\partial x} + X = a\frac{\partial^2 u}{\partial t} , \qquad (2.5.1)$$

$$\sigma = \frac{\partial u}{\partial x} - b\theta , \qquad (2.5.2)$$

$$\frac{\partial^2 \theta}{\partial x^2} = f\frac{\partial \theta}{\partial t} + g\frac{\partial^2 u}{\partial x\, \partial t} \qquad (2.5.3)$$

where

$$a = \left(\frac{l}{U\tau}\right)^2, \quad b = aT_0 , \quad f = \frac{\sigma c l^2}{k\tau} , \quad g = \frac{aEl^2}{k\tau} \qquad (2.5.4)$$

with $U = (E/\sigma)^{\frac{1}{2}}$, the velocity of elastic waves in the rod. It should be noted that b is independent of the choice of l and τ as also is

$$\varepsilon = \frac{bg}{f} = \frac{a^2ET}{\varrho c} \ . \tag{2.5.5}$$

If we are interested in low frequencies it is convenient to take l to be 1 cm. and τ to be 1 sec. The values of a, b, f, g and for four metals are shown in Table 5; here we have taken l = 1 cm., τ = 1 sec. and $T = 20°$ C.

TABLE 5

	Aluminium	Copper	Iron	Lead
a	4.029×10^{-12}	7.539×10^{-12}	5.984×10^{-12}	7.063×10^{-11}
b	7.62×10^{-3}	4.98×10^{-3}	1.026×10^{-2}	9.67×10^{-3}
f	1.168	0.899	5.208	4.152
g	0.860	0.479	3.536	1.470
$\varepsilon = bg/f$	5.61×10^{-3}	2.65×10^{-3}	6.97×10^{-3}	3.42×10^{-3}

For high frequencies it is more appropriate to take

$$\tau = \frac{1}{\omega^*} \ , \quad l = \frac{U}{\omega^*} \ , \tag{2.5.6}$$

where U denotes the elastic wave velocity and ω^* denotes the frequency

$$\omega^* = \frac{\varrho c}{k} U^2 \ . \tag{2.5.7}$$

For this choice of units we have

$$a = 1, \quad b = aT_0, \quad f = 1, \quad g = \frac{aE}{\varrho c}. \qquad (2.5.8)$$

TABLE 6

	Aluminium	Copper	Iron	Lead
U(km./sec.)	5.09	3.52	5.21	1.19
ω^*(sec.$^{-1}$)	3.03×10^{11}	1.11×10^{11}	1.43×10^{12}	5.88×10^{11}
l(cm.)	1.68×10^{-6}	3.16×10^{-6}	3.64×10^{-7}	2.02×10^{-7}
b	7.62×10^{-3}	4.98×10^{-3}	1.03×10^{-2}	9.67×10^{-3}
g	0.736	0.533	0.679	0.354
ω_c(sec.$^{-1}$)	7.90×10^{13}	6.10×10^{13}	8.94×10^{13}	2.05×10^{13}

The values of ω^* and l for aluminium, copper, iron and lead are given in Table 6. Again it is assumed that T_0 = 20°C. The range of frequencies actually obtainable in a solid is limited above by the cut-off frequency ω_c of the Debye spectrum. For longitudinal vibrations in a bar

$$\omega_c = 2\pi U(3\sigma/4\pi M)^{1/3}. \qquad (2.5.9)$$

The values of T_0 are included in Table 6.

In practical problems still another system of units may be employed. For instance, if we are considering the propagation of stress in a rod of length 1 metre, then it is obviously desirable to take $l = 10^2$ cm. and we may choose

$$\tau = 10^2/U.. \qquad (2.5.10)$$

With this choice of units we find that we may write the equations in the forms

$$\frac{\partial \sigma}{\partial x} = \frac{\partial^2 u}{\partial t^2}, \tag{2.5.11}$$

$$\sigma = \frac{\partial u}{\partial x} - b\theta , \tag{2.5.12}$$

$$\xi \frac{\partial^2 \theta}{\partial x^2} = \frac{\partial \theta}{\partial t} + \frac{g}{f} \frac{\partial^2 u}{\partial x \partial t}, \tag{2.5.13}$$

where

$$\xi = \frac{1}{f} = \frac{k}{\varrho c U} \times 10^{-2}, \quad \frac{g}{f} = \frac{aE}{\varrho c}, \quad b = aT \tag{2.5.14}$$

all the physical quantities being measured in c.g.s. units. For the metals we have been considering we get the values of ξ and τ given in Table 7;

TABLE 7

	Aluminium	Copper	Iron	Lead
τ(sec.)	1.97×10^{-4}	2.84×10^{-4}	1.92×10^{-4}	8.40×10^{-4}
ξ	1.68×10^{-8}	3.16×10^{-8}	3.68×10^{-9}	2.02×10^{-8}

in this system of units g/f has the same value as g has in Table 6, and b has the same value as in Tables 5 and 6 (on the assumption that $T_0 = 20°C.$).

Equation (2.5.13) governs the flow of heat in the

rod when the surface of the rod is rendered impervious to heat. The equation has to be modified when radiation takes place into a medium at constant temperature, T_0 say.

This means that equation (2.1.13) must be replaced by

$$\frac{\partial^2\theta}{\partial x^2} = f\frac{\partial\theta}{\partial t} + g\frac{\partial^2 u}{\partial x\partial t} + j\theta , \qquad (2.5.15)$$

where f and g have the same values as before and

$$j = \frac{Hpl^2}{Ak} . \qquad (2.5.16)$$

We shall consider first of all the propagation of waves in a rod whose surface is impervious to heat. If we put $X=0$ in equation (2.1.5) and put each of the quantities u, σ, θ proportional to $e^{i\omega t}$, then equations (2.5.1), (2.5.2) and (2.5.3) become

$$D\sigma = -a\omega^2 u , \qquad (2.5.17)$$

$$\sigma = Du - b\theta , \qquad (2.5.18)$$

$$D^2\theta = i\omega f\theta + i\omega g Du , \qquad (2.5.19)$$

where $D = d/dx$. It follows that each of the independent variables satisfies the equation

$$\begin{vmatrix} a\omega^2 & D & 0 \\ D & -1 & b \\ i\omega g D & 0 & -(D^2 - i\omega f) \end{vmatrix} \phi = 0 ,$$

which may be written in the form

$$(D^2-\mu_1^2)(D^2-\mu_2^2)\phi = 0 , \qquad (2.5.20)$$

where μ_1^2, μ_2^2 are the roots of the quadratic equation

$$(\mu^2 + a\omega^2)(\mu^2 - i\omega f) - i\omega b g \mu^2 = 0 . \qquad (2.5.21)$$

If we write

$$\Delta^2 = \left[a\omega^2 - i(f+bg)\omega\right]^2 + 4ia f\omega^3 , \qquad (2.5.22)$$

then we have

$$\mu_1^2 = \frac{1}{2}\left[-a\omega^2 + i\omega(f+bg) + \Delta\right] , \qquad (2.5.23)$$

$$\mu_2^2 = \frac{1}{2}\left[-a\omega^2 + i\omega(f+bg) - \Delta\right] \qquad (2.5.24)$$

and the linearly independent solutions of the set of equations (2.5.17), (2.5.18) and (2.5.19) therefore have solutions of the form

$$\begin{bmatrix} u \\ \sigma \\ \theta \end{bmatrix} = \begin{bmatrix} u_1 \\ \sigma_1 \\ \theta_1 \end{bmatrix} e^{\mu_1 x + i\omega t} + \begin{bmatrix} u_2 \\ \sigma_2 \\ \theta_2 \end{bmatrix} e^{-\mu_1 x + i\omega t} + \begin{bmatrix} u_3 \\ \sigma_3 \\ \theta_3 \end{bmatrix} e^{\mu_2 x + i\omega t} + \begin{bmatrix} u_4 \\ \sigma_4 \\ \theta_4 \end{bmatrix} e^{-\mu_2 x + i\omega t}$$

where the u_i, σ_i, θ_i $(i = 1,2,3,4)$ are constants. Of these twelve constants only four can be chosen arbitrarily; the remaining eight must be chosen in such a way that the equations (2.5.17), (2.5.18) and (2.5.19) are satisfied identically.

It is obvious from an examination of equations (2.5.23) and (2.5.24) that μ_2 is the root which corresponds to the longitudinal elastic wave. Now the phase velocity of the wave $\phi = \phi_0 \exp(-\mu_2 x + i\omega t)$ is given by the equation

$$V = \frac{\omega}{\mathrm{Im}(\mu_2)} \tag{2.5.25}$$

and the attenuation constant is

$$q = \mathrm{Re}(\mu_2) \; . \tag{2.5.26}$$

In problems of this kind the most appropriate units are those of Table 6. Putting $a = f = 1$, $bg = a^2 ET/\varrho c = \varepsilon$ in equations (2.5.23) and (2.5.24) we then find that

$$\mu_1^2 = \frac{1}{2}\left[-\omega^2 + i(1+\varepsilon)\omega + \Delta\right], \quad \mu_2^2 = \frac{1}{2}\left[-\omega^2 + i(1+\varepsilon)\omega - \Delta\right], \tag{2.5.27}$$

where

$$\Delta^2 = \left[\omega^2 - i(1+\varepsilon)\right]^2 + 4i\omega^3 . \tag{2.5.28}$$

Remembering that, in these units, the unit of velocity is U and the unit of length is U/ω^* we find from equations (2.5.25), (2.5.26) that

$$V = \frac{\omega U}{\mathrm{Im}(\mu_2)}, \quad q = \mathrm{Re}(\mu_2)\frac{U}{\omega^*} , \tag{2.5.29}$$

where μ_2 is given by the second of equations (2.5.27).

The algebraic expressions for the roots μ_1, μ_2 would be very cumbersome in the general case but it is a simple matter to approximate to them if ω is either very small or very large in comparison with unity (*i.e.* in comparison with ω^*). If $\omega \ll 1$ in this system of units then it is easily shown that μ_1 and μ_2 take the approximate values $\mu_1^{(0)}$, $\mu_2^{(0)}$ defined by the equations

$$\mu_1^{(0)} = \pm\left(\frac{1}{2}\omega\right)^{\frac{1}{2}}\left(1+\frac{1}{2}\varepsilon\right)(1+i)\,, \quad \mu_2^{(0)} = \pm\left\{\frac{1}{2}\varepsilon\left(1-\frac{5}{2}\varepsilon\right)+\left(1-\frac{1}{2}\varepsilon\right)i\omega\right\}\,, \tag{2.5.30}$$

where it has been assumed that ε also is small. Similarly if $\omega \gg 1$ in this system it can be shown that μ_1 and μ_2 take the approximate values $\mu_1^{(\infty)}, \mu_2^{(\infty)}$ where

$$\mu_1^{(\infty)} = \pm\left(\frac{1}{2}\omega\right)^{\frac{1}{2}}\left\{\left(1-\frac{\varepsilon}{2\omega}\right)+i\left(1+\frac{\varepsilon}{2\omega}\right)\right\}, \quad \mu_2^{(\infty)} = \pm\left(\frac{1}{2}\varepsilon+i\omega\right). \tag{2.5.31}$$

It follows from equations (2.5.30), (2.5.31) and (2.5.29) that in "ordinary" units

$$q = \begin{cases} \left(1-\frac{5}{2}\varepsilon\right)q_\infty(\omega/\omega^*)^2\,, & \omega \ll \omega^*\,, \\ q_\infty\,, & \omega \gg \omega^*\,, \end{cases} \tag{2.5.32}$$

where

$$q_\infty = \frac{1}{2}\varepsilon\frac{U}{\omega^*} \tag{2.5.33}$$

and from equation (2.5.29) that for $\omega \ll \omega^*$

$$V = \left(1+\frac{1}{2}\varepsilon\right)U\,, \tag{2.5.34}$$

while for $\omega \gg \omega^*$, $V = U$.

The attenuation constant q is therefore an increasing function of the frequency ω of the waves varying like ω^2 for low frequencies and approaching the value q_∞ asymptotically as $\omega \to \infty$. Equation (2.5.34) shows that for low frequencies, *i.e.* for frequencies less than 10^{10} sec.$^{-1}$, the phase velocity of the longitudinal elastic waves is $\left(1+\frac{1}{2}\varepsilon\right)$ times that of the elastic wave in a medium not exhibiting a thermal effect. This result can be interpreted in a different way. If measurements of the velocity of longitudinal waves in rods were used to determine the Young's modulus of the material of the rods then the "dynamical" value E_d of the Young's modulus would be related to the statical value E through the equation

$$E_d = (1+\varepsilon)E \ .$$

From the values given in Table 5 we see that in the case of aluminium the dynamical value will be 0.6 per cent higher than the statical value, for copper it will be 0.3 per cent higher, for iron 0.7 per cent higher and for lead 0.3 per cent higher. The effect of the thermal properties of the bar on these values is therefore rather slight.

In order to test the validity of these arguments a series of calculations based on equations (2.5.23), (2.5.24), (2.5.25), (2.5.26), was carried out for the four metals listed in Table 5 for a range of frequencies from $10^{-11}\omega^*$ to $10^{3}\omega^*$. Some of the results are shown in Tables 8, 9. It will be observed from Table 8 that if $\omega/\omega^* = 10^{-2}$ then $V = \left(1+\frac{1}{2}\varepsilon\right)U$; further

TABLE 8

ω/ω^*	V/U			
	Aluminium	Copper	Iron	Lead
10^{-2}	1.0028	1.0013	1.0035	1.0017
10^{-1}	1.0028	1.0013	1.0034	1.0017
1	1.0014	1.0007	1.0018	1.0009
10	1.0000	1.0000	1.0000	1.0000
10^2	1.0000	1.0000	1.0000	1.0000
$1+\frac{1}{2}\varepsilon$	1.0028	1.0013	1.0035	1.0017

TABLE 9

ω/ω^*	q/q_∞			
	Aluminium	Copper	Iron	Lead
10^{-2}	0.9846×10^{-4}	0.9917×10^{-4}	0.9815×10^{-4}	0.9910×10^{-4}
10^{-1}	0.9863×10^{-2}	0.9837×10^{-2}	0.9729×10^{-2}	0.9828×10^{-2}
1	0.4991	0.4996	0.4988	0.5000
10	0.9899	0.9901	0.9896	0.9910
10^2	0.9995	0.9999	0.9993	1.0000
$1-\frac{5}{2}\varepsilon$	0.9860	0.9933	0.9826	0.9914

calculations confirm that over the range $10^{-11} \ll \omega/\omega^* \ll 10^{-2}$ the phase velocity of longitudinal elastic waves in the rod has this constant value. When $\omega = \omega^*$ the phase velocity of these waves is almost exactly $\left(1+\frac{1}{4}\varepsilon\right)U$ and when $\omega \gg 10\omega^*$ it falls to the value U, independent of ε . Table 9 shows the values of q/q_∞ for the same range of frequencies; for higher values of ω it was found that $q = q_\infty$ while for $\omega < 10^{-2}\omega^*$ it was found that q was given accurately by equation (2.5.32).

It is obvious from these results that there is a sharp change in the values of V and q in the vicinity of the frequency ω^*. It would seem therefore that the frequency ω^*, which was introduced for the purely mathematical reason that its choice put the basic equations in a simple form, has in fact a definite physical significance.

2.6 Vibrations of a Circular Cylinder

We now study small vibrations of an infinite circular cylinder. Since, as we have shown above, the propagation of elastic shear waves is uninfluenced by the thermal properties of the solid, we need only consider longitudinal vibrations.

In the representation (1.5.5) we replace Φ by χ and take, in cylindrical coordinates (ϱ, ϕ, z)

$$\vec{A} = (0, \Psi, 0).$$

Assuming that both χ and Ψ are functions of ϱ and z only and making use of the results of § 1.8 *(i)* we obtain the displacement field

$$u_\varrho = \frac{\partial \chi}{\partial \varrho} - \frac{\partial \Psi}{\partial z}, \quad u_\phi = 0, \quad u_z = \frac{\partial \chi}{\partial z} + \frac{1}{\varrho}\frac{\partial}{\partial \varrho}(\varrho \Psi). \qquad (2.6.1)$$

If we make use of the notation

$$\Delta_a = \frac{\partial^2}{\partial \varrho^2} + \frac{1}{\varrho}\frac{\partial}{\partial \varrho} + \frac{\partial^2}{\partial z^2} \qquad (2.6.2)$$

we find that the thermoelastic equations become

$$V_T^2 \Delta_a \chi - (\alpha/\sigma\varkappa)\theta = \frac{\partial^2 \chi}{\partial t^2} \tag{2.6.3}$$

$$k\Delta_a \theta = \sigma c_\varepsilon \frac{\partial \theta}{\partial t} + (\alpha T_0/\varkappa)\frac{\partial}{\partial t}\Delta_a \phi \tag{2.6.4}$$

$$V_S^2 \Delta_a \Psi = \frac{\partial^2 \Psi}{\partial t^2} . \tag{2.6.5}$$

Assuming a solution of these equations of the form

$$(\chi, \Psi, \theta) = \{X(\varrho), \Psi(\varrho), \Theta(\varrho)\} \exp\{i(qz - \omega t)\} \tag{2.6.6}$$

we find that the functions $X(\varrho), \Psi(\varrho), \Theta(\varrho)$ satisfy the system of second order ordinary differential equations

$$\{\mathcal{B}_0 + (\omega^2/V_T^2 - q^2)\}X = (\alpha/\sigma\varkappa V_T^2)\Theta$$

$$\{\mathcal{B}_0 + (i\omega\sigma c_\varepsilon/k - q^2)\}\Theta = -(i\omega\alpha T_0/k\varkappa)({}_0 - q^2)X$$

$$\{\mathcal{B}_1 + (\omega^2/V_S^2 - q^2\}\Psi = 0$$

where

$$\mathcal{B}_\nu = \frac{\partial^2}{\partial \varrho^2} + \frac{1}{\varrho}\frac{\partial}{\partial \varrho} - \frac{\nu^2}{\varrho^2} . \tag{2.6.7}$$

If we assume that u_ϱ, u_z and Θ each remain finite along the z-axis we find that

$$\theta = (\sigma\varkappa V_T^2/\alpha)\left[(\omega^2/V_T^2-\zeta_1^2)AJ_0(\lambda_1\varrho)+(\omega^2/V_T^2-\zeta_2^2)BJ_0(\lambda_2\varrho)\right] \quad (2.6.8)$$

$$X = AJ_0(\lambda_1\varrho)+BJ_0(\lambda_2\varrho) \quad (2.6.9)$$

$$\Psi = CJ_1(\lambda_3\varrho) \quad (2.6.10)$$

where in terms of ζ_1^2, ζ_2^2 and ζ_3^2 the roots of equations (2.4.8) and (2.4.9)

$$\lambda_j^2 = \zeta_j^2 - q^2, \quad (j=1,2,3) \quad (2.6.11)$$

and A, B and C are arbitrary constants.

If the surface of the cylinder $\varrho = a$ is free from stress

$$\sigma_{\varrho\varrho}(a,z) = \sigma_{\varrho z}(a,z) = 0 \quad (2.6.12)$$

for all real values of z, and if there is radiation from the surface of the cylinder (into a medium maintained at the reference temperature) we have the condition

$$\frac{\partial\theta(a,z)}{\partial\varrho} + h\theta(a,z) = 0 . \quad (2.6.13)$$

From the stress-strain relations we find that the conditions (2.6.12) are equivalent to

$$\sigma V_T^2\Delta_a\chi - 2\sigma V_S^2(\varrho^{-1}\chi_\varrho + \chi_{zz} + \Psi_{\varrho z}) = (\alpha/\varkappa)\theta , \quad (\varrho = a), \quad (2.6.14)$$

$$2\chi_{\varrho z} + \mathcal{B}_1 \Psi - \Psi_{zz} = 0 , \quad (\varrho = a) . \qquad (2.6.15)$$

Substituting from equations (2.6.6), (2.6.8), (2.6.9) and (2.6.10) into equations (2.6.13) - (2.6.15) we find that

$$\{hJ_0(\lambda_1 a) - \lambda_1 J_1(\lambda_1 a)\}(\omega^2/V_T^2 - q^2 - \lambda_1^2)A$$

$$+ \{hJ_0(\lambda_2 a) - \lambda_2 J_1(\lambda_2 a)\}(\omega^2/V_T^2 - q^2 - \lambda_2^2)B = 0 ,$$

$$\{(2q^2 - \omega^2/V_S^2)a\, J_0(\lambda_1 a) + 2\lambda_1 J_1(\lambda_1 a)\} A$$

$$+ \{(2q^2 - \omega^2/V_S^2)a\, J_0(\lambda_2 a) + 2\lambda_2 J_1(\lambda_2 a)\} B =$$

$$= 2iqa\, \{\lambda_3 a\, J_0(\lambda_3 a) - J_1(\lambda_3 a)\} C ,$$

$$q\lambda_1 J_1(\lambda_1 a) A + q\lambda_2 J_1(\lambda_2 a) B = (q^2 - \omega^2/2V_S^2) J_1(\lambda_3 a) .$$

Eliminating A , B and C from these equations we obtain the equation

$$(2q^2 - \omega^2/V_S^2) f + \lambda_1 \lambda_2 (\lambda_1^2 - \lambda_2^2) J_1(\lambda_1 a) J_1(\lambda_2 a) g$$
$$= h(2q^2 - \omega^2/V_S^2)(\lambda_1^2 - \lambda_2^2) J_0(\lambda_1 a) J_1(\lambda_2 a) - hfg , \qquad (2.6.16)$$

where

$$f = \lambda_1(q^2 + \lambda_1^2 - \omega^2/V_T^2) J_1(\lambda_1 a) J_0(\lambda_2 a) - \lambda_2(q^2 + \lambda_2^2 - \omega^2/V_T^2) J_0(\lambda_1 a) J_1(\lambda_2 a) , \qquad (2.6.17)$$

$$g = 4(q^2/a) \left[\lambda_3 a\, J_0(\lambda_3 a)/J_1(\lambda_3 a) - \omega^2/V_S^2\right] . \qquad (2.6.18)$$

The wave number q is determined as a function of the frequency by the equations (2.4.8), (2.4.9), (2.6.11) and (2.6.16).

CHAPTER 3

COUPLED PROBLEMS OF THERMOELASTICITY

3.1 Thermal Stresses in Thin Metallic Rods

We shall consider first the state of stress and the distribution of temperature in the semi-infinite rod $x \gg 0$ when the end $x = 0$ is disturbed by the application of stresses or temperatures which have a time dependence of the sinusoidal form $e^{i\omega t}$ where ω is real. (Cf. Sneddon, 1958).

If we assume that the amplitude of the thermal and elastic waves do not increase indefinitely as $x \to \infty$ then we must take values of μ_1, μ_2 with negative real parts. It is readily verified that if μ_1 and μ_2 are given by equations (2.5.27), negative real parts being selected, then the expressions

$$u(x,t) = \frac{\mu_1 b}{\mu_1^2 + a\omega^2} C_1 e^{\mu_1 x + i\omega t} + \frac{\mu_2 b}{\mu_2^2 + a\omega^2} C_2 e^{\mu_2 x + i\omega t}, \qquad (3.1.1)$$

$$\sigma(x,t) = -b a \omega^2 \left[\frac{C_1}{\mu_1^2 + a\omega^2} e^{\mu_1 x + i\omega t} + \frac{C_2}{\mu_2^2 + a\omega^2} e^{\mu_2 x + i\omega t} \right] \qquad (3.1.2)$$

$$\theta(x,t) = C_1 e^{\mu_1 x + i\omega t} + C_2 e^{\mu_2 x + i\omega t}, \qquad (3.1.3)$$

satisfy the basic equations (2.5.1) to (2.5.3) with $X = 0$. These solutions lead to the boundary values

$$u(0,t) = \left[\frac{\mu_1 bC_1}{\mu_1^2 + a\omega^2} + \frac{\mu_2 bC_2}{\mu_2^2 + a\omega^2}\right]e^{i\omega t}, \qquad (3.1.4)$$

$$\sigma(0,t) = -ba\omega^2\left[\frac{C_1}{\mu_1^2 + a\omega^2} - \frac{C_2}{\mu_2^2 + a\omega^2}\right]e^{i\omega t}, \qquad (3.1.5)$$

$$\theta(0,t) = (C_1 + C_2)e^{i\omega t}, \qquad (3.1.6)$$

$$\frac{\partial\theta(0,t)}{\partial x} = (\mu_1 C_1 + \mu_2 C_2)e^{i\omega t}. \qquad (3.1.7)$$

Case (i):

$$\theta(0,t) = \Theta e^{i\omega t}, \quad \sigma(0,t) = 0 .$$

For this set of boundary conditions we must choose C_1 and C_2 so that

$$C_1 + C_2 = \Theta, \quad (\mu_2^2 + a\omega^2)C_1 + (\mu_1^2 + a\omega^2)C_2 = 0,$$

from which it follows that

$$C_1 = \frac{\mu_1^2 + a\omega^2}{\mu_2^2 - \mu_1^2}\Theta, \quad C_2 = \frac{\mu_2^2 + a\omega^2}{\mu_2^2 - \mu_1^2}\Theta,$$

so that the desired solution is

$$u(x,t) = -\frac{b\Theta}{\mu_2^2 - \mu_1^2}(\mu_1 e^{\mu_1 x + i\omega t} - \mu_2 e^{\mu_2 x + i\omega t}), \qquad (3.1.8)$$

$$\sigma(x,t) = \frac{a\,b\omega^2\Theta}{\mu_2^2-\mu_1^2}\left(e^{\mu_1 x+i\omega t}-e^{\mu_2 x+i\omega t}\right), \tag{3.1.9}$$

$$\theta(x,t) = -\frac{\Theta}{\mu_2^2-\mu_1^2}\left[(\mu_1^2+a\,\omega^2)e^{\mu_1 x+i\omega t}-(\mu_2^2+a\,\omega^2)e^{\mu_2 x+i\omega t}\right]. \tag{3.1.10}$$

In particular

$$u(0,t) = \frac{b\Theta}{\mu_1+\mu_2}e^{i\omega t}. \tag{3.1.11}$$

Case (ii):

$$\theta(0,t) = \Theta e^{i\omega t}, \quad u(0,t) = 0.$$

In a similar way we can show that the solution of this boundary value problem is

$$u(x,t) = -\frac{\mu_1\mu_2\Theta}{(\mu_2-\mu_1)(\mu_1\mu_2-a\,\omega^2)}\left(e^{\mu_1 x+i\omega t}-e^{\mu_2 x+i\omega t}\right), \tag{3.1.12}$$

$$\sigma(x,t) = \frac{b\,a\,\omega^2\Theta}{(\mu_2-\mu_1)(\mu_1\mu_2-a\,\omega^2)}\left(\mu_2 e^{\mu_1 x+i\omega t}-\mu_1 e^{\mu_2 x+i\omega t}\right), \tag{3.1.13}$$

$$\theta(x,t) = -\left\{\mu_2(\mu_1^2+a\,\omega^2)e^{\mu_1 x+i\omega t}-\mu_1(\mu_2^2+a\,\omega^2)e^{\mu_2 x+i\omega t}\right\}$$

$$\frac{\Theta}{(\mu_2-\mu_1)(\mu_1\mu_2-a\,\omega^2)}. \tag{3.1.14}$$

In particular

$$\sigma(0,t) = \frac{b\,a\,\omega^2\theta}{\mu_1\mu_2 - a\,\omega^2}\,e^{i\omega t}\,.$$

Since $\mu_1\mu_2 = -i\alpha f\mu^3$ it follows that in this case

$$\sigma(0,t) = b\theta\left(\frac{a\,\omega}{2f}\right)^{\frac{1}{2}}(1+i)e^{i\omega t}\,. \qquad (3.1.15)$$

Case (iii):

$$\sigma(\theta,t) = \Pi e^{i\omega t}\,, \quad \partial\theta(0,t)/\partial x = 0\,.$$

In this instance the constants C_1 and C_2 are determined by the equations

$$\mu_1 C_1 + \mu_2 C_2 = 0\,, \qquad \frac{C_1}{\mu_1^2 + a\,\omega^2} - \frac{C_2}{\mu_2^2 + a\,\omega^2} = \frac{\Pi}{b\,a\,\omega^2}$$

so that

$$\frac{C_1}{\mu_2} = -\frac{C_2}{\mu_1} = \frac{\Pi(\mu_1^2 + a\,\omega^2)(\mu_2^2 + a\,\omega^2)}{b\,a\,\omega(\mu_1 - \mu_2)(\mu_1^2 + \mu_1\mu_2 + \mu_2^2 + a\,\omega^2)}\,. \qquad (3.1.16)$$

The solution of the boundary value problem is given by equations (3.1.1), (3.1.2) and (3.1.3) with C_1 and C_2 given by the equations (3.1.16). In particular we have the boundary expressions

$$\theta(0,t) = -\Pi\,\frac{(\mu_1^2 + a\,\omega^2)(\mu_2^2 + a\,\omega^2)}{b\,a\,\omega^2(\mu_1^2 + \mu_1\mu_2 + \mu_2^2 + a\,\omega^2)}\,e^{i\omega t}\,, \qquad (3.1.17)$$

$$u(0,t) = -\Pi \frac{\mu_1\mu_2(\mu_1+\mu_2)}{a\,\omega^2(\mu_1^2+\mu\,\mu+\mu_2^2+a\,\omega^2)}\, e^{i\omega t} \,. \tag{3.1.18}$$

Case (iv):

$$\sigma(0,t) = \Pi e^{i\omega t}, \quad \theta(0,t) = 0 \,.$$

The solution of this problem is

$$u(x,t) = \frac{\Pi}{a\,\omega^2(\mu_1^2-\mu_2^2)} \{\mu_1(\mu_2^2+a\,\omega^2)e^{\mu_1 x+i\omega t} - \mu_2(\mu_1^2+a\,\omega^2)e^{\mu_2 x+i\omega t}\}\,, \tag{3.1.19}$$

$$\sigma(x,t) = \frac{\Pi}{(\mu_1^2-\mu_2^2)} \{(\mu_2^2+a\,\omega^2)e^{\mu_1 x+i\omega t} - (\mu_1^2+a\,\omega^2)e^{\mu_2 x+i\omega t}\}\,, \tag{3.1.20}$$

$$\theta(x,t) = \frac{\Pi(\mu_1^2+a\,\omega^2)(\mu_2^2+a\,\omega^2)}{b a\,\omega^2(\mu_1^2-\mu_2^2)} (e^{\mu_1 x+i\omega t} - e^{\mu_2 x+i\omega t})\,, \tag{3.1.21}$$

whence

$$u(0,t) = -\frac{(\mu_1\mu_2 - a\,\omega^2)\Pi e^{i\omega t}}{a\,\omega^2(\mu_1+\mu_2)} \tag{3.1.22}$$

Case (v):

$$u(0,t) = A e^{i\omega t}, \quad \partial\theta/\partial x = 0 \,.$$

The solution of this boundary value problem is

$$u(x,t) = \frac{A}{\mu_2^2 - \mu_1^2}\{(\mu_2^2 + a\omega^2)e^{\mu_1 x + i\omega t} - (\mu_1^2 + a\omega^2)e^{\mu_2 x + i\omega t}\}, \quad (3.1.23)$$

$$\sigma(x,t) = \frac{a\omega^2 A}{\mu_2^2 - \mu_1^2}\left\{\frac{\mu_2^2 + a\omega^2}{\mu_1}e^{\mu_1 x + i\omega t} - \frac{\mu_1^2 + a\omega^2}{\mu_2}e^{\mu_2 x + i\omega t}\right\}, \quad (3.1.24)$$

$$\theta(x,t) = \frac{A(\mu_1^2 + a\omega^2)(\mu_2^2 + a\omega^2)}{b(\mu_2^2 - \mu_1^2)}(\mu_1^{-1}e^{\mu_1 x + i\omega t}\ \mu_2^{-1}e^{\mu_2 x + i\omega t}) \quad (3.1.25)$$

from which we have

$$\sigma(0,t) = -\frac{a\omega^2 A(\mu_1^2 + \mu_1\mu_2 + \mu_2^2 + a\omega^2)}{\mu_1\mu_2(\mu_1 + \mu_2)}e^{i\omega t}, \quad (3.1.26)$$

$$\theta(0,t) = \frac{A(\mu_1^2 + a\omega^2)(\mu_2^2 + a\omega^2)}{b\mu_1\mu_2(\mu_1 + \mu_2)}e^{i\omega t}. \quad (3.1.27)$$

Case (vi):

$$u(0,t) = Ae^{i\omega t}, \quad \theta(0,t) = 0.$$

In this case we have

$$u(x,t) = \frac{A}{(\mu_2 - \mu_1)(\mu_1\mu_2 - a\omega^2)}\{\mu_1(\mu_2^2 + a\omega^2)e^{\mu_1 x + i\omega t} - \mu_2(\mu_1^2 + a\omega^2)e^{\mu_2 x + i\omega t}\} \quad (3.1.28)$$

$$\sigma(x,t) = -\frac{a\omega^2 A}{(\mu_2-\mu_1)(\mu_1\mu_2 - a\omega^2)}\{(\mu_2^2 + a\omega^2)e^{\mu_1 x + i\omega t} - (\mu_1^2 + a\omega^2)e^{\mu_2 x + i\omega t}\}, \tag{3.1.29}$$

$$\theta(x,t) = \frac{A(\mu_1^2 + a\omega^2)(\mu_2^2 + a\omega^2)}{b(\mu_2-\mu_1)(\mu_1\mu_2 - a\omega^2)}(e^{\mu_1 x + i\omega t} - e^{\mu_2 x + i\omega t}), \tag{3.1.30}$$

and hence

$$\sigma(0,t) = -\frac{a\omega^2(\mu_1+\mu_2)Ae^{i\omega t}}{\mu_1\mu_2 - a\omega^2}. \tag{3.1.31}$$

The reciprocal nature of the pair of equations (3.1.31) and (3.1.32) will be observed; the pair (3.1.18) and (3.1.26) are similarly related.

We shall consider use of the method of the Laplace transform to study the conditions in the semi-inifinite rod $x \geqslant 0$ (in the absence of body forces) when the end $x = 0$ is subjected to certain stresses and thermal conditions. In addition to assuming that $X = 0$ we assume that at $t = 0$ the stress and displacement are zero as is the deviation θ of the temperature from the reference temperature.

If we take the Laplace transforms of both sides of each of the equations (3.1.1) to (3.1.3) we obtain the ordinary differential equations

$$D\bar{\sigma} = a s^2 \bar{u} \tag{3.1.32}$$

$$\bar{\sigma} = D\bar{u} - b\bar{\theta} \tag{3.1.33}$$

$$D^2\bar{\theta} = sf\bar{\theta} + gsD\bar{u} \tag{3.1.34}$$

for the Laplace transforms (with respect to the time)

$$\begin{aligned} \bar{u}(x,s) &= \mathcal{L}[u(x,t); t \longrightarrow s] \\ \bar{\sigma}(x,s) &= \mathcal{L}[\sigma(x,t); t \longrightarrow s] \\ \bar{\theta}(x,s) &= \mathcal{L}[\theta(x,t); t \longrightarrow s] \end{aligned} \tag{3.1.35}$$

of u, σ and θ. It is readily seen that each of these quantities satisfies the fourth order ordinary differential equation.

$$(D^2 - \varkappa_1^2)(D^2 - \varkappa_2^2)\bar{\phi} = 0 , \tag{3.1.36}$$

where $D = \partial/\partial x$ and

$$\varkappa_1^2 + \varkappa_2^2 = a s^2 + (f + bg)s , \quad \varkappa_1^2 \varkappa_2^2 = a f s^3 . \tag{3.1.37}$$

On the assumption that σ, u and θ all tend to zero (or at worst remain finite) as $x \rightarrow \infty$ we take solutions of equations (3.1.32), (3.1.33) and (3.1.34) of the form

$$\bar{u}(x,s) = -b\left[\frac{\varkappa_1 C_1 e^{-\varkappa_1 x}}{\varkappa_1^2 - a s^2} + \frac{\varkappa_2 C_2 e^{-\varkappa_2 x}}{\varkappa_2^2 - a s^2}\right] , \tag{3.1.38}$$

$$\bar{\sigma}(x,s) = a\,bs^2\left[\frac{C_1 e^{-\varkappa_1 x}}{\varkappa_1^2 - a s^2} + \frac{C_2 e^{-\varkappa_2 x}}{\varkappa_2^2 - a s^2}\right], \qquad (3.1.39)$$

$$\bar{\Theta}(x,s) = C_1 e^{-\varkappa_1 x} + C_2 e^{-\varkappa_2 x},$$

where it is assumed that $Re\varkappa_1$ and $Re\varkappa_2$ are both positive, and C_1 and C_2 are determined from the conditions at $x = 0$. To facilitate the calculation of C_1 and C_2 from the boundary conditions we have the relations

$$\bar{u}(0,s) = -b\left[\frac{\varkappa_1 C_1}{\varkappa_1^2 - a s^2} + \frac{\varkappa_2 C_2}{\varkappa_2^2 - a s^2}\right], \qquad (3.1.40)$$

$$\bar{\sigma}(0,s) = a\,bs^2\left[\frac{C_1}{\varkappa_1^2 - a s^2} + \frac{C_2}{\varkappa_2^2 - a s^2}\right], \qquad (3.1.41)$$

$$\bar{\Theta}(0,s) = C_1 + C_2, \qquad (3.1.42)$$

$$D\bar{\Theta}(0,s) = -\varkappa_1 C_1 - \varkappa_2 C_2. \qquad (3.1.43)$$

It is obvious from this set of equations that if the boundary values of any two of the four quantities $\bar{u}$, $\bar{\sigma}$, $\bar{\Theta}$, $D\bar{\Theta}$ are prescribed those of the remaining two are uniquely determinded.

For instance, suppose that the stress is prescribed to have the value $\Pi(t)$ on the free end and it is assumed that there is no flux of heat across this end. Then $\sigma(0,t) = \Pi(t)$ and $\partial\theta(0,t)/\partial x = 0$. From these conditions it follows immediately that $D\bar{\theta}(0,s) = 0$ and that $\bar{\sigma}(0,s) = \bar{\Pi}(s)$, where $\bar{\Pi}(s)$ is the Laplace transform of the prescribed function $\Pi(t)$. Substituting these values in equations (3.1.41) and (3.1.43) and solving for C_1 and C_2 we find that

$$\frac{C_1}{-x_2} = \frac{C_2}{x_1} = \frac{g\bar{\Pi}(s)}{(x_2-x_1)(f+bg)+a^{1/2}f^{1/2}s^{1/2}} .$$

Substituting these values in equation (3.1.42) we obtain the relation

$$\bar{\theta}(0,s) = -\frac{g\bar{\Pi}(s)}{(f+bg)+a^{1/2}f^{1/2}s^{1/2}} .$$

If we choose $1/\omega^*$ as our unit of time and U/ω^* as our unit of length then $a = f = 1$ and $bg = \varepsilon$ so that we have

$$\bar{\theta}(0,s) = -\frac{g\Pi(s)}{s^{1/2}+\alpha}, \qquad \alpha = 1+\varepsilon . \tag{3.1.44}$$

Now if we write

$$F(\alpha,t) = \pi^{-1/2}t^{-1/2} - \alpha e^{\alpha^2 t}\,\mathrm{Erfc}(\alpha t^{1/2}) \tag{3.1.45}$$

we find that

$$\bar{F}(\alpha,s) = \frac{1}{s^{1/2}+\alpha}$$

so that making use of the convolution theorem for Laplace transforms (A.27) we obtain the equation

$$\bar{\theta}(0,t) = -g\int_0^t \Pi(t')F(1+\varepsilon, t-t')dt' \qquad (3.1.46)$$

by means of which to determine the variation of temperature at the free end (x=0) of the rod.

For example, if in coventional units

$$\Pi(t) = \begin{cases} -\mu p_0, & 0 \leq t \leq t_1 \\ 0, & t > t_1, \end{cases}$$

so that in our present units we have

$$\Pi(t) = \begin{cases} -p_0, & 0 \leq t \leq t_1\omega^* \\ 0, & t > t_1\omega^* \end{cases}$$

and it can be shown that

$$\theta(0,t) = \begin{cases} \frac{p_0 g}{\alpha}\left[1-e^{\alpha^2 t}\mathrm{Erfc}(\alpha t^{1/2})\right], & 0 \leq t \leq t_1\omega^*, \\ \frac{p_0 g}{\alpha}\left[e^{\alpha^2(t-t_1\omega^*)}\mathrm{Erfc}\{\alpha(t-t_1\omega^*)^{1/2}\}-e^{\alpha^2 t}\mathrm{Erfc}(\alpha t^{1/2})\right], & t>t_1\omega^*. \end{cases}$$

Returning to conventional units we have

$$\theta(0,t) = \begin{cases} p_0\theta_m\{1-e^{\gamma^2}\mathrm{Erfc}(\gamma t^{1/2})\}, & 0 << t << t_1, \\ p_0\theta_m\{e^{\gamma^2(t-t_1)}\mathrm{Erfc}\,\gamma(t-t_1)^{1/2}-e^{\gamma^2}\mathrm{Erfc}(\gamma t^{1/2})\}, & t >> t_1, \end{cases} \qquad (3.1.47)$$

where

$$\Theta_m = \frac{gT}{1+\varepsilon}, \quad \gamma^2 = (1+\varepsilon)^2 \omega^* . \tag{3.1.48}$$

Values of Θ_m and γ for the metals we have considered previously are given in Table 1 for a reference temperature of 20°C.

TABLE 1

	Aluminium	Copper	Iron	Lead
Θ_m(°K)	214.4	155.6	197.5	103.4
γ(sec$^{-1/2}$)	5.53×10^{5}	3.34×10^{5}	1.20×10^{6}	7.69×10^{5}
$\pi^{-1/2}\Theta_{m/\gamma}$(°Ksec$^{1/2}$)	2.19×10^{-4}	2.63×10^{-4}	9.26×10^{-5}	7.58×10^{-5}

Since γ is very large we may use the asymptotic expansion

$$e^{x^2}\mathrm{Erfc}(x) \sim \frac{1}{\sqrt{\pi}x}\left(1-\frac{1}{2x^2}\right) \tag{3.1.49}$$

to obtain the formulae

$$\theta(0,t) = \begin{cases} p_0\Theta_m\{1-\gamma^{-1}(\pi t)^{-1/2}\} & 0 \le t < t_1 \\ p_0\Theta_m & t = t_1 \\ \frac{p_0\Theta_m}{\gamma\sqrt{\pi}}(t-t_1)^{1/2-}-t^{-1/2} & t > t_1 . \end{cases} \tag{3.1.50}$$

Furthermore if $t \gg t_1$ we have the approximate formula

$$\theta(0,t) = \frac{p_0\Theta_m}{\gamma(\pi t_1)^{1/2}}\left(\frac{t_1}{t}\right)^{3/2} . \tag{3.1.51}$$

Similar methods may be employed to determine the state of stress and the distribution of temperature in the finite rod $0 \ll x \ll l$. There is a variety of thermoelastic problems associated with such a finite rod, the solution of each depending on the boundary conditions imposed.

In the first instance consider the vibrations possible in the rod when the ends $x = 0$ and $x = l$ are subjected to the conditions

$$\sigma = 0 , \quad \frac{\partial \theta}{\partial x} = 0 . \tag{3.1.52}$$

It is readily verified that if μ_1 and μ_2 are given by equations (2.5.27), then the expression

$$\sigma = \sigma_1 \sinh \mu_1 x + \sigma_2 \sinh \mu_2 x + \sigma_3 (\cosh \mu_1 x - \cosh \mu_2 x) e^{i\omega t} , \tag{3.1.53}$$

in which $\sigma_1 , \sigma_2 , \sigma_3$ are constants satisfies the equation (2.5.20) and the condition that $\sigma = 0$ when $x = 0$. From the relation

$$a b \omega^2 \theta = -(D^2 + a \omega^2) \sigma$$

we deduce that the corresponding expression for θ is given by the equation

$$\theta = -\frac{1}{a b \omega^2} \{\sigma_1 (\mu_1^2 + a \omega^2) \sinh \mu_1 x + \sigma_2 (\mu_2^2 + a \omega^2) \sinh \mu_2 x$$
$$+ \sigma_3 [(\mu_1^2 + a \omega^2) \cosh \mu_1 x - (\mu_2^2 + a \omega^2) \cosh \mu_2 x]\} e^{i\omega t} . \tag{3.1.54}$$

It follows immediately that the expressions (3.1.53) and (3.1.54) satisfy the boundary conditions (3.1.52) at $x = 0$ and $x = l$ provided that

$$\mu_1(\mu_1^2 + a\omega^2)\sigma_1 + \mu_2(\mu_2^2 + a\omega^2)\sigma_2 = 0 ,$$

$$\mu_1(\mu_1^2 + a\omega^2)\sigma_1 \cosh\mu_1 l + \mu_2(\mu_2^2 + a\omega^2)\sigma_2 \cosh\mu_2 l$$

$$+ [\mu_1(\mu_1^2 + a\omega^2)\sinh\mu_1 l - \mu_2(\mu_2^2 + a\omega^2)\sinh\mu_2 l]\sigma_3 = 0 ,$$

$$\sigma_1 \sinh\mu_1 l + \sigma_2 \sinh\mu_2 l + \sigma(\cosh\mu_1 l - \cosh\mu_2 l) = 0 .$$

Eliminating the constants σ_1, σ_2, σ_3 from these equations we see that solutions of the type (3.1.53) and (3.1.54) are possible if ω is a root of the transcendental equation

$$\begin{vmatrix} \mu_1(\mu_1^2 + a\omega^2) & \mu_2(\mu_2^2 + a\omega^2) & 0 \\ [\mu_1(\mu_1^2 + a\omega^2)\cosh\mu_1 l] & [\mu_2(\mu_2^2 + a\omega^2)\cosh\mu_2 l] & [\mu_1(\mu_1^2 + a\omega^2)\sinh\mu_1 l - \mu_2(\mu_2^2 + a\omega^2)\sinh\mu_2 l] \\ \sinh\mu_1 l & \sinh\mu_2 l & \cosh\mu_1 l - \cosh\mu_2 l \end{vmatrix} = 0,$$

where μ_1 and μ_2 are defined in terms of ω by the equations (2.5.27). Expanding the determinant on the left-hand side of this equation we find that the frequency equation becomes

$$\begin{aligned} &2\mu_1\mu_2(\mu_1^2 + a\omega^2)(\mu_2^2 + a\omega^2)(\cosh\mu_1 l \cosh\mu_2 l - 1) \\ &= [\mu_1^2(\mu_1^2 + a\omega^2)^2 + \mu_2^2(\mu_2^2 + a\omega^2)^2]\sinh\mu_1 l \sinh\mu_2 l . \end{aligned} \tag{3.1.55}$$

The method of the Laplace transform can also be applied to problems concerning rods of finite length l. Suppose, for instance, that we have the following boundary conditions for the stress $\sigma(x,t)$ and the temperature variation $\theta(x,t)$:

$$\sigma(0,t) = \Pi(t)\,, \quad \frac{\partial\theta(0,t)}{\partial x} = 0$$
$$\sigma(l,t) = 0\,, \quad \frac{\partial\theta(l,t)}{\partial x} = 0 \tag{3.1.56}$$

then we may take a solution of equation (3.1.36) in the form

$$\bar{\sigma}(x,s) = A\sinh\varkappa_1(l-x) + B\sinh\varkappa_2(l-x) + C\cosh\varkappa_1(l-x) + D\cosh\varkappa_2(l-x)\,,$$

where $\varkappa_1, \varkappa_2$ are defined by the equations (3.1.37). The corresponding expression for the Laplace transform of the temperature variation is

$$-abs^2\bar{\theta}(x,s) = (\varkappa_1^2 - as^2)A\sinh\varkappa_1(l-x) + C\cosh\varkappa_1(l-x) + (\varkappa_2^2 - as^2)B\sinh\varkappa_2(l-x) + D\cosh\varkappa_2(l-x)\,.$$

If these forms are to satisfy the boundary conditions (3.1.56) we must choose A, B, C, D so that they satisfy the equations

$$A\sinh\varkappa_1 l + B\sinh\varkappa_2 l + C\cosh\varkappa_1 l + D\cosh\varkappa_2 l = \bar{\Pi}(s)\,,$$

$$C + D = 0\,,$$

$$\varkappa_1(\varkappa_1^2 - as^2)(A\cosh\varkappa_1 l + C\sinh\varkappa_1 l) + \varkappa_2(\varkappa_2^2 - as^2)(B\cosh\varkappa_2 l + D\sinh\varkappa_2) = 0\,,$$

$$\varkappa_1(\varkappa_1^2 - a s^2)A + \varkappa_2(\varkappa_2^2 - a s^2)B = 0 \ .$$

Solving these equations we find that

$$A = \varkappa_2(\varkappa_2^2 - a s^2)\Phi, \quad B = -\varkappa_1(\varkappa_1^2 - a s^2)\Phi, \quad C = \Psi, \quad D = -\Psi$$

where

$$\Phi = \varkappa_1(\varkappa_1^2 - a s^2)\sinh \varkappa_1 l - \varkappa_2(\varkappa_2^2 - a s^2)\sinh \varkappa_2 l \ \frac{\bar{\Pi}(s)}{\Delta(s)} ,$$

$$\Psi = -\varkappa_1 \varkappa_2(\varkappa_1^2 - a s^2)(\varkappa_2^2 - a s^2)(\cosh \varkappa_1 l - \cosh \varkappa_2 l) \ \frac{\bar{\Pi}(s)}{\Delta(s)}$$

with

$$\Delta(s) = 2\varkappa_1 \varkappa_2(\varkappa^2 - a s^2)(\varkappa^2 - a s^2)(\cosh \varkappa_1 l \cosh \varkappa_2 l - 1)$$

$$- \left[\varkappa_1^2(\varkappa_1^2 - a s^2)^2 + \varkappa_2^2(\varkappa_2^2 - a s^2)\right]\sinh \varkappa_1 l \sinh \varkappa_2 l \ .$$

The relation of the function $\Delta(s)$ defined by equation (3.1.57) to the frequency equation (3.1.55) is obvious.

3.2 Approximate Solutions in the One-Dimensional Case

In this section we shall consider a method of obtaining approximate solutions to the complete set of equations (2.5.1) to (2.5.3).

In some cases it is a simple matter to find a set of functions (σ_0 , u_0 , θ_0) satisfying the boundary conditions of the problem and the "unlinked" equations

$$\frac{\partial \sigma_0}{\partial x} = a\frac{\partial^2 u_0}{\partial t^2} , \tag{3.2.1}$$

$$\sigma_0 = \frac{\partial u_0}{\partial x} , \tag{3.2.2}$$

$$\frac{\partial^2 \theta_0}{\partial x^2} = f\frac{\partial \theta_0}{\partial t} . \tag{3.2.3}$$

From these solutions we may determine an approximate solution of the set (2.5.1) to (2.5.3). Let

$$(u,\sigma,\theta) = \sum_{r=0}^{\infty}(u_r,\sigma_r,\theta_r)g^r , \tag{3.2.4}$$

where $(\sigma_0, u_0, \theta_0)$ satisfy equations (3.2.1) to (3.2.3) then, substituting in equations (2.5.1) to (2.5.3) we find that $(u_0, \sigma_0, \theta_0)$ satisfy the equations

$$\left(\frac{\partial^2}{\partial x^2} - a\frac{\partial^2}{\partial t^2}\right)u_r = \left(\frac{b}{g}\right)\frac{\partial \theta_{r-1}}{\partial x} , \tag{3.2.5}$$

$$\left(\frac{\partial^2}{\partial x^2} - f\frac{\partial}{\partial t}\right)\theta_r = \frac{\partial u_{r-1}}{\partial x} , \tag{3.2.6}$$

$$\sigma_r = \frac{\partial u_r}{\partial x} - \left(\frac{b}{g}\right)\theta_{r-1} \tag{3.2.7}$$

$(r > 1)$, and the relevant physical quantities vanish on the boundary of the rod.

As an example of this method, suppose that we wish to find the approximation solution of the problem considered in case *(i)* of §3.1. In this case the approximate solutions of equations (3.2.1), (3.2.2) and (3.2.3) would be

$$\theta_0 = \Theta e^{i\omega t-(1+i)(1/2\, f\omega)^{1/2} x}, \qquad u_0 = \sigma_0 = 0 .$$

We therefore have

$$\left(\frac{\partial^2}{\partial x^2} - a\frac{\partial^2}{\partial t^2}\right)u_1 = -(1+i)\left(\frac{1}{2}f\omega\right)^{1/2}\frac{b}{g}\Theta e^{i\omega t-(1+i)(1/2\, f\omega)^{1/2} x}$$

which has solution

$$u_1 = Ae^{i\omega(t-a^{1/2}x)} - \frac{(b\Theta/g)\left(\frac{1}{2}f\omega\right)^{1/2}(1+i)e^{i\omega t-(1+i)(1/2\, f\omega)^{1/2} x}}{a\omega^2+if\omega} .$$

Inserting this expression in equation (3.2.7) we find that

$$\sigma_1(x,t) = -ia^{1/2}\omega Ae^{i\omega(t-a^{1/2}x)} - \frac{(b\Theta/g)a\omega^2 e^{i\omega t-(1+i)(1/2\, f\omega)^{1/2} x}}{a\omega^2+if\omega}$$

and the boundary conditions demand that $\sigma_1(0,t) = 0$ so that we have

$$A = \frac{ia^{1/2}\omega(b\Theta/g)}{a\omega^2+if\omega} .$$

We therefore obtain the approximate solution

$$\tilde{u}(x,t) = g u_1(x,t)$$
$$= \frac{i b a^{1/2} \omega \Theta}{a\omega^2 + i f \omega} e^{i\omega(t - a^{1/2}x)} - \frac{b\left(\frac{1}{2} f\omega\right)^{1/2}(1+i)\Theta}{a\omega^2 + i f\omega} e^{-(1+i)(1/2\, f\omega)^{1/2} x + i\omega t} .$$

For this solution we have the boundary value

$$\tilde{u}(0,t) = -\left(\frac{2}{f\omega}\right)^{1/2} b\Theta \frac{1 - i\left[1 + (2a\omega/f)^{1/2}\right]}{1 + \left[1 + (2a\omega/f)^{1/2}\right]^2} e^{i\omega t} .$$

Comparing this solution with the exact solution (3.1.11) we find that

$$\varkappa(\omega) = \frac{|\tilde{u}(0,t)|}{|u(0,t)|} = \left(\frac{2}{f\omega}\right)^{1/2} \frac{|\mu_1 + \mu_2|}{\left[1 + \left\{1 + (2a\omega/f)^{1/2}\right\}^2\right]^{1/2}} .$$

To test the validity of this approximation the values of $\varkappa(\omega)$ for aluminium were examined over a wide range of frequencies. The results of this numerical investigation are shown in Table 2. It will be observed that over a very wide frequency

TABLE 2

ω/ω^*	10^{-10}	10^{-8}	10^{-6}	10^{-4}	10^{-2}	1	10^2
$\varkappa(\omega)$	1.0028	1.0028	1.0028	1.0028	1.0024	1.0006	1.0000

range the function $\varkappa(\omega)$ does not differ from unity by more than 0.0028 showing the approximate solution obtained by this crude method is accurate to within about one quarter per cent.

3.3 The Stresses Produced in an Infinite Elastic Solid by Uneven Heating

We shall now consider the distribution of stress in an infinite elastic body containing heat sources, i.e. we shall consider the solution of equations (1.5.3), (1.5.4) for $-\infty < x_1, x_2, x_3 < \infty$ and known function Q. It will be assumed that there are no body forces acting in the solid. We follow Eason and Sneddon (1958).

(a) General Theory.

It is convenient to write the equations (1.5.3), (1.5.4) in dimensionless form. (Sneddon and Berry, 1958, p. 124). If we take a typical length l as our unit of length, a time τ as our unit of time, the reference temperature T_0 as the unit of temperature, and the rigidity modulus μ as unit of stress we find that the field equations (1.3.4), (1.4.4), (1.5.1) respectively take the dimensionless forms

$$\sigma_{pq,q} + F_p = a\frac{\partial^2 u_p}{\partial t^2}, \tag{3.3.1}$$

$$\sigma_{pq} = [(\beta^2 - 2)\vartheta - b\theta]\delta_{pq} + 2\varepsilon_{pq}, \tag{3.3.2}$$

$$\Delta_3\theta + \Theta = f\frac{\partial\theta}{\partial t} + g\frac{\partial\vartheta}{\partial t}, \tag{3.3.3}$$

where β is the ratio $(2+\lambda/\mu)^{1/2}$,

$$F_p = \frac{l\sigma}{\mu}X_p. \quad \Theta = \frac{Ql^2}{kT_0} \tag{3.3.4}$$

define new source functions and

$$a = \left(\frac{l}{V_s \tau}\right)^2, \quad b = \frac{\gamma T_0}{\mu}, \quad f = \frac{l^2 \sigma c_\varepsilon}{k\tau}, \quad g = \frac{\gamma l^2}{k\tau} \qquad (3.3.5)$$

where $V_S = (\mu/\sigma)^{1/2}$ is the velocity of shear waves and $\gamma = \alpha(3\lambda + 2\mu)$.

The values of the constants a, b, f and g occurring in these equations will depend upon the choice of the basic units of time and length. In some problems it may be convenient to choose l to be 1 cm. and τ to be 1 sec. The values of a, b, f and g for four common metals and this choice of l and τ are shown in Table 1; here we have taken T to be 293°K. For other choices of l and τ the corresponding values of a, b, f and g can be de-

TABLE 1

Values of the constants a, b, f and g for four common metals with $l = 1$cm., $\tau = 1$ sec. and $T = 293$ °K.

	Aluminium	Copper	Iron	Lead
a	1.034×10^{-11}	2.166×10^{-11}	1.532×10^{-11}	2.034×10^{-10}
b	0.0639	0.0417	0.0089	0.2320
f	1.168	0.899	5.208	4.152
g	2.687	1.497	8.035	12.25
$\varepsilon = bg/f\beta^2$	3.56×10^{-2}	1.68×10^{-2}	2.97×10^{-4}	7.33×10^{-2}

rived from equations (3.3.5). It will be observed in these equations that b is independent of the choice of l and τ as also are g/f and

$$\varepsilon = \frac{bg}{\beta^2 f} = \frac{\gamma^2 T}{\varrho c \beta^2}.$$

For theoretical investigations at high frequencies it is desirable to choose the units introduced in § 3.2.1.

With this choice of fundamental units

$$a = \beta^2, \quad b = \frac{\gamma T}{\mu}, \quad f = 1, \quad g = \frac{\gamma}{\varrho c},$$

where $\beta = V_P/V_S$, the ratio of the velocity of pure P-waves to that of S-waves in the solid. Thus

$$a = \beta^2 = \frac{\lambda + 2\mu}{\mu} = \frac{2(1-\eta)}{1-2\eta}$$

where η denotes Poisson's ratio of the solid. The values of b are the same as those given in Table 1 and the value of g is merely that of g/f obtained from Table 1. The values of ε are unaltered. As we have said, this system of units is of value in some investigations but it should be borne in mind that ω^* is, for most materials, much higher than the highest frequency obtainable by using ultrasonic techniques and only about one-hundredth of the cut-off frequency ω_c of the Debye spectrum. For instance, for iron, $\omega^* = 1.75 \times 10^{12}$ sec.$^{-1}$ while $\omega_c = 9.95 \times 10^{13}$ sec^{-1}; this means that the unit of length is 3.31×10^{-7} cm. so that the system is not of much interest in the discussion of engineering problems except possibly for very high frequency phenomena.

In practical problems still another system of units may be employed. For instance, if we are considering the propagation of thermal stress in a plate of thickness 1 metre, it is desirable to take $l = 10^2$ cm. and we may choose

$$\tau = (10^2/V_S)\text{sec.},$$

where the velocity of shear waves, V_2 is expressed in cm. per sec. With this choice of units we may write the equations of thermoelasticity in the forms

$$\sigma_{pq,q} + F_p = \frac{\partial^2 u_p}{\partial t^2} \tag{3.3.1*}$$

$$\sigma_{pq} = (\beta^2 - 2) - b\theta\delta_{pq} + 2\varepsilon_{pq} \tag{3.3.2*}$$

$$\varkappa(\Delta_3\theta + \theta) = \frac{\partial\theta}{\partial t} + \frac{g}{b}\cdot\frac{\partial\Delta}{\partial t} \tag{3.3.3*}$$

where

$$\varkappa = f^{-1} = \kappa V_S^{-1} 10^{-2}, \quad \frac{g}{f} = \frac{\gamma}{\varrho c}, \quad b = \frac{\gamma T}{\mu}$$

all the physical quantities being measured in c.g.s. units. For the metals considered in Table 1 we get the values of $\varkappa$ and τ shown in Table 2; in this system of units b has the same set of values as in Table 1 and the values of g/f can be obtained by dividing the fourth row of Table 1 by the third row. In calculating τ and $\varkappa$ in Table 2 we have again assumed that $T_0 = 293°$K.

TABLE 2

	Aluminium	Copper	Iron	Lead
τ(secs.)	3.215×10^{-4}	4.654×10^{-4}	3.072×10^{-4}	1.427×10^{-3}
$\varkappa$	2.750×10^{-8}	5.172×10^{-8}	5.887×10^{-9}	3.429×10^{-8}

To solve the set of partial differential equations (3.3.1) to (3.3.3) we introduce the four-dimensional transform of each of the physical quantities occurring in the basic equations. We use the notation

$$\hat{f}(\xi_1, \xi_2, \xi_3, \omega) = \mathcal{F}_{(4)}\left[f(\underset{\sim}{x}, t); \underset{\sim}{x} \rightarrow \underset{\sim}{\xi}, t \rightarrow \omega\right].$$

If we multiply both sides of equations (3.3.1), (3.3.2) and (3.3.3) by

$$\exp\{i(\xi_p x_p + \omega t)\},$$

integrate over E_4 and make use of the analogue, for n-dimensional transforms, of equation (A.1) we find that these partial differential equations are equivalent to the set of algebraic equations

$$-i\xi_q \hat{\sigma}_{pq} = -a\omega^2 \hat{u}_p, \tag{3.3.6}$$

$$\hat{\sigma}_{pq} = -i(\beta^2 - 2)\xi_r \hat{u}_r + b\hat{\theta}\,\delta_{pq} - i(\xi_q \hat{u}_p + \xi_p \hat{u}_q), \tag{3.3.7}$$

$$-\xi^2 \hat{\theta} + \hat{\Theta} = -if\omega\hat{\theta} - g\omega\xi_q \hat{u}_q, \tag{3.3.8}$$

where $\xi^2 = \xi_q \xi_q$, from which we may obtain expressions for the Fourier transforms of the temperature and of the components of the stress tensor and the displacement vector in terms of the Fourier transform of the source function Θ.

From equation (3.3.8) we find that

$$\hat{\theta} = \frac{g\omega\xi_q\hat{u}_q}{\xi^2 - if\omega} + \frac{\hat{\Theta}}{\xi^2 - if\omega} \tag{3.3.9}$$

If we substitute from this equation into equation (3.3.7) and insert the resulting expression for $\hat{\sigma}_{pq}$ into equation (3.3.6) we obtain the equation

$$\hat{u}_p = \frac{ib\xi_p\hat{\Theta}}{(\xi^2 - if\omega)(\beta^2\xi^2 - a\omega^2) - ibg\omega\xi^2} \, . \tag{3.3.10}$$

Substituting this expression back in equation (3.3.9) we find that Θ is given by the equation

$$\hat{\theta} = \frac{(\beta^2\xi^2 - a\omega^2)\hat{\Theta}}{(\xi^2 - if\omega)(\beta^2\xi^2 - a\omega^2) - ibg\omega\xi^2} \, ,$$

which may be written in the form

$$\hat{\theta} - \hat{\theta}^c = \frac{ibg\omega\xi^2\hat{\Theta}}{(\xi^2 - if\omega)(\xi^2 - a\omega^2)(\beta^2\xi^2 - a\omega^2) - ibg\omega\xi^2} \tag{3.3.11}$$

where

$$\hat{\theta}^c = \frac{\hat{\Theta}}{\xi^2 - if\omega} \tag{3.3.12}$$

denotes the Fourier transform of the "classical" solution. Substituting from equation (3.3.10) and (3.3.11) into (3.3.7) we obtain the expression

$$\hat{\sigma}_{pq} = -b\hat{\Theta}^c \delta_{pq} + \frac{b[(\beta^2-2)\xi^2\delta_{pq} + 2\xi_p\xi_q]\hat{\Theta} - ib^2 g\omega\xi^2(\xi^2 - if\omega)^{-1}\hat{\Theta}\delta_{pq}}{(\xi^2 - if\omega)(\beta^2\xi^2 - a\omega^2) - ibg\omega\xi^2} \tag{3.3.13}$$

for the Fourier transforms of the components of the stress tensor. It is also easily shown from equation (3.3.10) that the Fourier transform of the dilatation is given by the equation

$$\hat{\Theta} = \frac{b\hat{\Theta}\xi}{(\xi^2 - if\omega)(\beta^2\xi^2 - a\omega^2) - ibg\omega\xi^2} . \tag{3.3.14}$$

Using the Fourier incersion theorem we deduce from equation (3.3.10) that

$$u_p = \frac{b}{4\pi^2}\int_{W_4} \frac{i\xi_p\hat{\Theta}\exp\{-i(\xi_p x_p + \omega t)\}\,dW}{(\xi^2 - if\omega)(\beta^2\xi^2 - a\omega^2) - ibg\omega\xi^2} \tag{3.3.15}$$

where $dW = d\xi_1 d\xi_2 d\xi_3 d\omega$ and W_4 is the entire $\xi_1\xi_2\xi_3\omega$ -space. Similarly from equation (3.3.13) we deduce the formula

$$\sigma_{pq} = -b\Theta^c\delta_{pq}$$
$$+ \frac{b}{4\pi^2}\int_{W_4} \frac{\{(\beta^2-2)\xi^2\delta_{pq} + 2\xi_p\xi_q - i\omega bg\xi^2(\xi^2 - if\omega)^{-1}\delta_{pq}\}\hat{\Theta}e^{-i(\xi_p x_p + \omega t)}}{(\xi^2 - if\omega)(\beta^2\xi^2 - a\omega^2) - ibg\omega\xi^2}\,dW \tag{3.3.16}$$

which determines the components of the stress tensor. The temperature within the solid is given by the equation

$$\Theta - \Theta^c = \frac{bg}{4\pi^2}\int_{W_4} \frac{i\omega\xi^2\hat{\Theta}\exp\{-i(\xi_p x_p + \omega t)\}dW}{[(\xi^2 - if\omega)(\beta^2\xi^2 - a\omega^2) - ibg\omega\xi^2](\xi^2 - if\omega)}, \tag{3.3.17}$$

where the "classical" temperature Θ^c has the form

$$\Theta^c = \frac{1}{4\pi^2}\int_{W_4} \frac{\hat{\Theta}}{(\xi^2 - if\omega)} \exp[-i(\xi_p x_p + \omega t)]dW . \tag{3.3.18}$$

(b) Plane Strain.

The solution appropriate to a two-dimensional problem of plane strain, in which the displacement vector at the point (x_1, x_2) has components (u_1, u_2) and the state of stress is uniquely determined by the three components $\sigma_{11}, \sigma_{22}, \sigma_{33}$ (cf. § 1.1.7), may be obtained from equations (3.3.15) and (3.3.16) by assuming that Θ is a function of x_1, x_2 and t only. We then find that the $\hat{\Theta}$ occurring in these equations should be replaced by the expression

$$(2\pi)^{1/2}\,\delta(\xi_3)\hat{\Theta}$$

where, now

$$\hat{\Theta}(\xi_1, \xi_2, \omega) = \mathfrak{F}_{(3)}[\Theta(x_1, x_2, t); (x_1, x_2, t) \to (\xi_1, \xi_2, \omega)] . \tag{3.3.19}$$

From equation (3.3.15) we deduce that the plane strain solution is

$$u_p = \frac{b}{(2\pi)^{3/2}}\int_{T_3} \frac{i\xi_p\hat{\Theta}(\xi_1, \xi_2, \omega)}{(\xi_1^2 + \xi_2^2 - if\omega)[(\xi_1^2 + \xi_2^2)\beta^2 - a\omega^2] - ibg\omega(\xi_1^2 + \xi_2^2)} e^{-ix} d\vec{T} \tag{3.3.20}$$

$(p=1,2)$ where $d\vec{T}=d\xi_1 d\xi_2 d\omega$, T_3 is the whole $\xi_1\xi_2\omega$-space and $\varkappa = x_1\xi_1 + x_2\xi_2 + \omega t$. The stress tensor is given by the tensor equation

$$\sigma_{pq} = \frac{b}{(2\pi)^{3/2}} \int_{T_3} \frac{\left[(\beta^2-2)(\xi_1^2+\xi_2^2)\delta_{pq} + 2\xi_p\xi_q - i\omega bg(\xi_1^2+\xi_2^2)(\xi_1^2+\xi_2^2-if\omega)^{-1}\delta_{pq}\right]}{(\xi_1^2+\xi_2^2-if\omega)\left[\beta^2(\xi_1^2+\xi_2^2)-a\omega^2\right]-i\omega bg(\xi_1^2+\xi_2^2)} \cdot$$

$$\cdot \hat{\Theta} e^{-i\varkappa} d\vec{T} - \frac{b}{(2\pi)^{3/2}}\delta_{pq} \int_{T_3} \frac{\hat{\Theta}}{(\xi_1^2+\xi_2^2-if\omega)} e^{-i\varkappa} d\vec{T} \qquad (3.3.21)$$

where $p,q=1,2$, and σ_{33} can be calculated by means of equation (2.7.7).

(c) Axial Symmetry.

In problems in which there is *axial symmetry* we choose the z-axis to be the axis of symmetry. The displacement vector then has components u_ϱ and u_z in the ϱ and z-directions respectively and zero components of stress which may be denoted in the Green-Zerna notation by $\sigma_{\varrho\varrho}, \sigma_{\phi\phi}, \sigma_{zz}, \sigma_{\varrho z}$ and which are related (in this system of coordinates) to u and u_z by the dimensionless equations

$$(\sigma_{\varrho\varrho}, \sigma_{\phi\phi}, \sigma_{zz}) = \left[(\beta^2-2)-b\Theta\right](1,1,1) + 2\left(\frac{\partial u_\varrho}{\partial \varrho}, \frac{u_\varrho}{\varrho}, \frac{\partial u_z}{\partial z}\right) \qquad (3.3.22)$$

and

$$\sigma_{\varrho z} = \frac{\partial u_\varrho}{\partial z} + \frac{\partial u_z}{\partial \varrho} \qquad (3.3.23)$$

where the dilatation ϑ takes the form

$$\vartheta = \frac{\partial u_\varrho}{\partial \varrho} + \frac{u_\varrho}{\varrho} + \frac{\partial u_z}{\partial z} \,. \tag{3.3.24}$$

The equations of motion (1.8.8) and (1.8.9) take the forms

$$\frac{\partial \sigma_{\varrho\varrho}}{\partial \varrho} + \frac{\partial \sigma_{\varrho z}}{\partial z} + \frac{\sigma_{\varrho\varrho} - \sigma_{\phi\phi}}{\varrho} + F_\varrho = a \frac{\partial^2 u_\varrho}{\partial t^2} \tag{3.3.25}$$

$$\frac{\partial \sigma_{\rho z}}{\partial \varrho} + \frac{\partial \sigma_{zz}}{\partial z} + \frac{\sigma_{\varrho z}}{\varrho} + F_z = a \frac{\partial^2 u_z}{\partial t^2} \tag{3.3.26}$$

and the heat conduction equation becomes

$$\left(\mathcal{B}_0 + \frac{\partial^2}{\partial z^2}\right)\theta + \Theta = f \frac{\partial \theta}{\partial t} + g \frac{\partial \Delta}{\partial t} \tag{3.3.27}$$

where

$$\mathcal{B}_\nu = \frac{\partial^2}{\partial \varrho^2} + \frac{1}{\varrho}\frac{\partial}{\partial \varrho} - \frac{\nu^2}{\varrho^2} \,.$$

Substituting from equations (3.3.22) into equations (3.3.25) and (3.3.26) we obtain the equations

$$\left(\beta^2 \mathcal{B}_1 + \frac{\partial}{\partial z^2}\right) u_\varrho + (\beta^2 - 1) \frac{\partial^2 u_z}{\partial \varrho \, \partial z} - b \frac{\partial \theta}{\partial \varrho} = a \frac{\partial^2 u_\varrho}{\partial t^2} \tag{3.3.28}$$

$$(\beta^2 - 1) \frac{\partial}{\partial z}\left(\frac{\partial u_\varrho}{\partial \varrho} + \frac{u_\varrho}{\varrho}\right) + \left(\mathcal{B}_0 + \beta^2 \frac{\partial^2}{\partial z^2}\right) u_z - b \frac{\partial \theta}{\partial z} = a \frac{\partial^2 u_z}{\partial t^2} \tag{3.3.29}$$

To solve these equations we introduce the integral transforms

$$\bar{u}_\varrho = \mathcal{F}_{(2)}[\mathcal{H}_1\{u_\varrho(\varrho, z, t); \varrho \to \xi\}; (z,t) \to (\zeta, \omega)] \tag{3.3.30}$$

$$\bar{u}_z(\xi, \zeta, \omega) = \mathcal{F}_{(2)}[\mathcal{H}_0\{u_z(\varrho, z, t); \varrho \to \xi\}; (z,t) \to (\zeta, \omega)] . \tag{3.3.31}$$

If we multiply both sides of equation (3.3.28) by

$$(2\pi)^{-1} \exp\{i(\zeta z + \omega t)\} \varrho J_1(\xi \varrho)$$

and both sides of equation (3.3.29) by

$$(2\pi)^{-1} \exp\{i(\zeta z + \omega t)\} \varrho J_0(\xi \varrho)$$

and, in both cases, integrate over the entire $\varrho z t$-space, we find, on making use of the properties of Fourier and Hankel transforms, that this pair of partial differential equations is equivalent to the pair of algebraic equations

$$(\beta^2 \xi^2 + \zeta^2 - a\omega^2)\bar{u} - i(\beta^2 - 1)\xi\zeta\bar{w} = b\xi\bar{\theta} , \tag{3.3.32}$$

$$i(\beta^2 - 1)\xi\zeta\bar{u} + (\xi^2 + \beta^2\zeta^2 - a\omega^2)\bar{w} = ib\zeta\bar{\theta} ,$$

where

$$\bar{\theta} = \mathcal{F}_{(2)}[\mathcal{H}_0\{\theta(\varrho, z, t); \varrho \to \xi\}; (z,t) \to (\omega, \zeta)] . \tag{3.3.33}$$

Solving these equations we find that

$$(\bar{u}, \bar{w}) = \frac{b\bar{\Theta}(\xi, i\zeta)}{\beta^2(\xi^2 + \zeta^2) - \alpha\omega^2} \tag{3.3.34}$$

If we multiply both sides of equation (3.3.27) by

$$(2\pi)^{-1}\exp\{i(\zeta z+\omega t)\}\varrho J_0(\xi\varrho)$$

and integrate over the whole $\varrho z t$ -space, we find that

$$(\xi^2+\zeta^2-if\omega)\bar{\theta}-i\omega g(\xi\bar{u}-i\zeta\bar{w})=\bar{\Theta}\,, \qquad (3.3.35)$$

where

$$\bar{\Theta}=\mathcal{F}_{(2)}\left[\mathcal{H}_0\{\Theta(\varrho,z,t);\,\varrho\to\xi\};\,(z,t)\to(\zeta,\omega)\right]. \qquad (3.3.36)$$

Solving the algebraic equations (3.3.32), (3.3.33) and (3.3.35) for the unknowns $\bar{u}$, $\bar{w}$, $\bar{\theta}$ in terms of the known quantity $\bar{\Theta}$ we find the expressions

$$\frac{\bar{u}}{b\xi}=\frac{\bar{w}}{ib\zeta}=\frac{\bar{\theta}}{\beta^2(\xi^2+\zeta^2)-a\omega^2}=\frac{\bar{\Theta}}{\beta^2(\xi^2+\zeta^2)-a\omega^2(\xi^2+\zeta^2-if\omega)-i\omega bg(\xi^2+\zeta^2)} \qquad (3.3.37)$$

for the transforms of the components of the displacement vector and of the temperature Θ. If we invert the equations (3.3.37) by the appropriate theorems for Fourier and Hankel transforms we find for the components of the displacement vector

$$u_\varrho=\frac{b}{2\pi}\int_{-\infty}^{\infty}\int_{-\infty}^{\infty}\exp[-i(\zeta z+\omega t)]\,d\zeta\,d\omega\int_0^{\infty}\frac{\xi^2\bar{\Theta}J_1(\xi\varrho)\,d\xi}{D(\xi,\zeta,\omega)}\,, \qquad (3.3.38)$$

$$u_z=\frac{ib}{2\pi}\int_{-\infty}^{\infty}\int_{-\infty}^{\infty}\exp[-i(\zeta z+\omega t)]\,d\zeta\,d\omega\int_0^{\infty}\frac{\xi\zeta\bar{\Theta}J_0(\xi\varrho)\,d\xi}{D(\xi,\zeta,\omega)}\,, \qquad (3.3.39)$$

and for the temperature variation

$$\Theta = \frac{1}{2\pi}\int_{-\infty}^{\infty}\int_{-\infty}^{\infty}\exp[-i(\zeta z+\omega t)]d\zeta d\omega\int_{0}^{\infty}\frac{\xi\{\beta^2(\xi^2+\zeta^2)-a\omega^2\}\bar{\Theta}J_0(\xi\varrho)d\xi}{D(\xi,\zeta,\omega)} \tag{3.3.40}$$

where

$$D(\xi,\zeta,\omega) = [\beta^2(\xi^2+\zeta^2) - a\omega^2](\xi^2+\zeta^2-if\omega)-i\omega bg(\xi^2+\zeta^2)\ . \tag{3.3.41}$$

(d) Quasi-Static Solution.

If we consider problems in which the c.g.s. system of units provides natural units of length and time, *i.e.* if the centimetre is the typical distance and the second the typical time then it is obvious from Table 1 that the constant a is very much smaller than the other constants b, g and f occurring in the dimensionless equations (3.3.1) to (3.3.3). It follows, by expanding the integrand in equation (3.3.15) in ascending powers of a, that the approximate solution

$$\begin{aligned} u_p = {} & \frac{b}{4\pi^2\beta^2}\int_{W_4}\frac{i\xi_p\hat{\Theta}\exp[-i(\xi_p x_p+\omega t)]dW}{\xi^2(\xi^2-if_1\omega)} \\ & + \frac{ab}{4\pi^2\beta^4}\int_{W_4}\frac{i\xi_p\omega^2(\xi^2-if\omega)\hat{\Theta}\exp[-i(\xi_p x_p+\omega t)]dW}{\xi^4(\xi^2-if_1\omega)^2} \end{aligned} \tag{3.3.42}$$

with

$$f_1 = f(1+\varepsilon) \tag{3.3.43}$$

will give a very accurate description of the displacement field in the elastic body. Because α is so very small we may in most cases take it to be zero and describe the displacement field by the quasi-static solution

$$u_p^{(0)} = \frac{b}{4\pi^2\beta^2}\int_{W_4} \frac{i\xi_p\hat{\Theta}\exp[-i(\xi_p x_p + \omega t)]dW}{\xi^2(\xi^2 - if_1\omega)^2}, \quad (p = 1,2,3). \tag{3.3.44}$$

Similarly equations (3.3.16) and (3.3.17) may be approximated to by the equations

$$\sigma_{pq}^{(0)} + b\Theta^c\delta_{pq} = \frac{b}{4\pi^2\beta^2}\int_{W_4} \frac{[(\beta^2-2)\xi^2\delta_{pq} + 2\xi_p\xi_q - i\omega bg\xi^2(\xi^2 - if\omega)^{-1}\delta_{pq}]\hat{\Theta}e^{-i(\xi_p x_p + \omega t)}}{\xi^2(\xi^2 - if_1\omega)}dW, \tag{3.3.45}$$

$$\Theta^{(0)} - \Theta^c = \frac{bg}{4\pi^2\beta^2}\int_{W_4} \frac{i\omega\xi^2\hat{\Theta}\exp[-i(\xi_p x_p + \omega t)]dW}{(\xi^2 - if_1\omega)(\xi^2 - if\omega)}, \tag{3.3.46}$$

where the "classical temperature" Θ^c has the form (3.3.18).

The corresponding approximation to the exact solution (3.3.20) of the two-dimensional problem is given by

$$u_p^{(0)} = \frac{b}{(2\pi)^{3/2}\beta^2}\int_{T_3} \frac{i\xi_p\hat{\Theta}\exp[-i(x_1\xi_1 + x_2\xi_2 + \omega t)]dT}{(\xi_1^2 + \xi_2^2)(\xi_1^2 + \xi_2^2 - if_2\omega)}, \quad (p = 1,2) \tag{3.3.47}$$

where, now, $\hat{\Theta}$ is defined by equation (3.3.19). For this solution the stress components are given by the equation

$$\sigma_{pq} = -\frac{b\delta_{pq}}{(2\pi)^{3/2}}\int_{T_3}\frac{\hat{\Theta}\exp[-i(x_1\xi_1+x_2\xi_2+\omega t)]d\vec{T}}{\xi_1^2+\xi_2^2-if\omega}$$

$$+\frac{b}{(2\pi)^{3/2}\beta^2}\int_{T_3}\frac{[(\beta^2-2)(\xi_1^2+\xi_2^2)\delta_{pq}+2\xi_p\xi_q-i\omega bg(\xi_1^2+\xi_2^2)(\xi_1^2+\xi_2^2-if\omega)^{-1}\delta_{pq}]}{(\xi_1^2+\xi_2^2)(\xi_1^2+\xi_2^2-if_1\omega)}$$

$$\times\hat{\Theta}\exp[-i(x_1\xi_1+x_2\xi_2+\omega t)]d\vec{T}\ . \tag{3.3.48}$$

The quasi-static solution of the axially symmetrical problem is found by putting $a = 0$ in equations (3.3.38) to (3.3.41). It is given by the equations

$$u_\varrho^{(0)} = \frac{b}{2\pi\beta^2}\int_{-\infty}^{\infty}\int_{-\infty}^{\infty}e^{-i(\zeta z+\omega t)}d\zeta d\omega\int_0^\infty\frac{\xi^2\hat{\Theta}J_1(\xi\varrho)d\xi}{(\xi^2+\zeta^2)(\xi^2+\zeta^2-if_1\omega)} \tag{3.3.49}$$

$$u_z^{(0)} = \frac{ib}{2\pi\beta^2}\int_{-\infty}^{\infty}\int_{-\infty}^{\infty}e^{-i(\zeta z+\omega t)}d\zeta d\omega\int_0^\infty\frac{\xi\zeta\hat{\Theta}J_0(\xi\varrho)d\xi}{(\xi^2+\zeta^2)(\xi^2+\zeta^2-if_1\omega)}\ , \tag{3.3.50}$$

$$\Theta^{(0)} = \frac{1}{2\pi}\int_{-\infty}^{\infty}\int_{-\infty}^{\infty}e^{-i(\zeta z+\omega t)}d\zeta d\omega\int_0^\infty\frac{\xi\hat{\Theta}J_0(\xi\varrho)d\xi}{\xi^2+\zeta^2-if_1\omega}\ . \tag{3.3.51}$$

The quantity $\hat{\Theta}$ occurring in these equations is defined by equation (3.3.36). It will be observed that in this case the temperature variation Θ has the same form as it has when it is governed by the simple equation for the conduction of heat; the only diffusion parameter f is replaced by the parameter f_1 defined by equation (3.3.43).

(e) Solution of Special Problems.

We shall now discuss the application of these general formulae to the solution of certain special problems.

(i) The Stress due to a Periodic Line Source.

We shall begin by considering the stress distribution arising from a line source of periodic strength which lies along the x_3-axis. If the source is of frequency Ω_0 then we have a two-dimensional problem in which

$$\Theta = F\delta(x_1)\delta(x_2)e^{i\Omega t} \tag{3.3.52}$$

where, in our units $\Omega = \Omega\tau$ and F is a constant. As a result of a simple integration we see that

$$\hat{\Theta} = \frac{F}{(2\pi)^{1/2}}\delta(\omega+\Omega) . \tag{3.3.53}$$

Substituting from equation (3.3.53) into equation (3.3.20) and performing the ω-integration we find that

$$u_\gamma = -\frac{bFx_\gamma}{4\pi^2\varrho}e^{i\Omega t}\frac{\partial I_1}{\partial\varrho} , \tag{3.3.54}$$

where $\varrho^2 = x_1 + x_2^2$ and

$$I_1 = \int_{-\infty}^{\infty}\int_{-\infty}^{\infty}\frac{e^{-i(\xi_1x_1+\xi_2x_2)}\,d\xi_1\,d\xi_2}{(\xi_1^2+\xi_2^2+if\Omega)[\beta^2(\xi_1^2+\xi_2^2)-a\Omega^2]+ibg\Omega(\xi_1^2+\xi_2^2)}$$

$$= 2\pi \int_0^\infty \frac{\lambda J_0(\varrho\lambda)d\lambda}{(\lambda^2 + if\Omega)(\beta^2\lambda^2 - a\Omega^2) + ibg\Omega\lambda^2} .$$

If ϱ_1^2 and ϱ_2^2 are the roots of the quadratic equation

$$\varrho^4 + (a\Omega^2/\beta^2 - if_1\Omega)\varrho^2 - iaf\Omega^3/\beta^2 = 0 \qquad (3.3.55)$$

in ϱ^2 then

$$u_p = \frac{bFx_p e^{i\Omega t}}{2\pi\beta^2\varrho} I_2 \qquad (p = 1,2) \qquad (3.3.56)$$

where

$$I_2 = \int_0^\infty \frac{\lambda^2 J_1(\varrho\lambda)d\lambda}{(\lambda^2 + \varrho_1^2)(\lambda^2 + \varrho^2)} \qquad (3.3.57)$$

$$= \frac{1}{\varrho_2^2 - \varrho_1^2}\left\{\int_0^\infty \frac{\lambda^2 J_1(\varrho\lambda)\,d\lambda}{\lambda^2 + \varrho_1^2} - \int_0^\infty \frac{\lambda^2 J_1(\varrho\lambda)d\lambda}{\lambda^2 + \varrho_2^2}\right\} . \qquad (3.3.58)$$

Similarly if we put

$$\Theta = \frac{F}{2\pi\varrho}\delta(\varrho)e^{i\Omega t} , \qquad (3.3.52a)$$

that is

$$\hat{\Theta} = F\delta(\omega + \Omega)\delta(\zeta) \qquad (3.3.53a)$$

in (3.3.39) we find that $u_z = 0$ and if we substitute this form in (3.3.38)

$$u_\varrho = \frac{bFe^{i\Omega t}}{2\pi\beta^2} I_2$$

in agreement with equation (3.3.46).

Using a well-known formula in the theory of Bessel functions (Watson 1944, p. 434) we find that

$$\int_0^\infty \frac{\lambda^2 J_1(\varrho\lambda)}{\lambda^2+k^2} d\lambda = kK_1(k\varrho), \qquad (3.3.59)$$

where $K_\nu(z)$ denotes the modified Bessel function of the second kind of order ν and argument z and, in terms of the first Hankel function,

$$K_\nu(z) = \frac{1}{2}\pi i e^{1/2\nu\pi i} H_\nu^{(1)}(iz)$$

(Watson 1944, p. 78). Inserting this expression for the integrals occurring in equations (3.3.48) we obtain the exact solution

$$u_\varrho = \frac{bFe^{i\Omega t}}{2\pi\beta^2(\varrho_2^2-\varrho_1^2)} \{\varrho_1 K_1(\varrho_1 r) - \varrho_2 K_1(\varrho_2 r)\}. \qquad (3.3.60)$$

The temperature variation in the solid is readily found from equation (3.3.40). We find that

$$\Theta = \frac{Fe^{i\Omega t}}{2\pi}\left[\frac{\varrho_1^2 + a\Omega^2/\beta^2}{\varrho_1^2-\varrho_2^2} I_3(\varrho_1) - \frac{\varrho_2^2 + a\Omega^2/\beta^2}{\varrho_1^2-\varrho_2^2} I_3(\varrho_2)\right],$$

where

$$I_3(\varrho_i) = \int_0^\infty \frac{\varrho J_0(\varrho r)d\varrho}{\varrho^2 + \varrho_i^2} = K_0(\varrho_i r)$$

(Watson 1944, p. 434). Hence we find that

$$\theta = \frac{Fe^{i\Omega t}}{2\pi}\left[\frac{\varrho_1^2 + a\Omega^2/\beta^2}{\varrho_1^2 - \varrho_2^2}K_0(\varrho_1 r) - \frac{\varrho_2^2 + a\Omega^2/\beta^2}{\varrho_1^2 - \varrho_2^2}K_0(\varrho_2 r)\right]. \quad (3.3.61)$$

The *quasi-static solution* obtained by inserting the value of $\bar{\theta}$ given by equation (3.3.53), (or (3.3.53a), into equation (3.3.47), (or equation (3.3.49)) may be put in the form

$$u_\varrho^{(0)} = \frac{bFe^{i\Omega t}I_4}{2\pi\beta^2},$$

where

$$I_4 = \int_0^\infty \frac{J_1(\varrho\lambda)d\lambda}{\lambda^2 + if_1\Omega}.$$

Making use of equation (3.3.59) and a well-known result in the theory of Bessel functions (Watson 1944, p.391) we find that

$$I_4 = \frac{1}{if_1\Omega}\int_0^\infty J_1(\lambda\varrho)\left[\frac{\lambda^2}{\lambda^2 + if_1\Omega} - 1\right]d\lambda = \frac{1}{if_1\Omega}\left[(f_1\Omega)^{1/2}e^{1/4\pi i}K_1(r_0 e^{1/4\pi i}) - \frac{1}{\varrho}\right],$$

where

$$r_0 = \varrho f_1^{1/2}\Omega^{1/2}. \quad (3.3.62)$$

Hence we have

$$u_{\varrho}^{(0)} = \frac{Fe^{i\Omega t}}{2\pi\beta^2 if_1\Omega\varrho}\left\{r_0e^{(1/4)\pi i}K_1(r_0e^{(1/4)\pi i})-1\right\} . \quad (3.3.63)$$

Furthermore, if we substitute from equation (3.3.53a) into equation (3.3.51) we find that the quasi-static value of the temperature is

$$\theta^{(0)} = \frac{Fe^{i\Omega t}}{2\pi}\int_0^\infty \frac{\xi J_0(\xi\varrho)d\xi}{\xi^2+if_1\Omega} . \quad (3.3.64)$$

Evaluating by the same formula as before (Watson 1944, p. 434) we find that

$$\theta^{(0)} = \frac{Fe^{i\Omega t}}{2\pi}K_0(r_0e^{(1/4)\pi i}) . \quad (3.3.65)$$

Using a known relation for modified Bessel functions of the second kind (Watson 1944, p. 80) we write this equation in the form

$$\theta^{(0)} = \frac{Fe^{i\Omega t}}{2\pi}\left\{K_0(r_0)-\frac{1}{4}\pi i I_0(r_0)\right\} .$$

From equations (3.3.61) and (3.3.65) we find that

$$\frac{\theta-\theta^{(0)}}{\theta^{(0)}} = \frac{(\varrho_1^2+a\Omega^2/\beta^2)K_0(\varrho_1\varrho)-(\varrho_2^2+a\Omega^2/\beta^2)K_0(\varrho_2\varrho)-(\varrho_1^2-\varrho_2^2)K_0(r_0e^{(1/4)\pi i})}{(\varrho_1^2-\varrho_2^2)K_0(r_0e^{(1/4)\pi i})} . \quad (3.3.66)$$

For small values of the parameter $(a\Omega/\beta^2 f)$ we can show that

$$\varrho_1 = (f_1\Omega)^{1/2} e^{1/4\pi i}\left\{1+\frac{i a\Omega}{2f\beta^2}\right\}, \quad \varrho_2 = \left(1-\frac{1}{2}\varepsilon\right)e^{1/2\pi i}\left(\frac{a^{1/2}\Omega}{\beta}\right)$$

from which it follows that

$$\frac{\varrho_1^2 + a\Omega^2/\beta^2}{\varrho_1^2-\varrho_2^2} = 1 - \frac{i\varepsilon a\Omega}{f_1\beta^2}, \quad \frac{\varrho_2^2 + a\Omega^2/\beta^2}{\varrho_1^2-\varrho_2^2} = -\frac{i\varepsilon a\Omega}{f_1\beta^2},$$

and

$$\varrho_1\varrho = r_0 e^{i(1/4\pi+\Psi)}, \quad \varrho_2\varrho = r_1 e^{1/2\pi i}, \quad \Psi = \varepsilon a\Omega/2f_1\beta^2,$$

where r_0 is defined by equation (3.3.62) and r_1 is defined by the equation

$$r_1 = \left(1-\frac{1}{2}\varepsilon\right)\frac{a^{1/2}\Omega}{\beta}\varrho . \qquad (3.3.67)$$

It should be observed that

$$\left(\frac{r_1}{r_0}\right)^2 = (1-\varepsilon)\frac{a\Omega}{\beta^2 f_1}$$

so that, if $a\Omega/\beta^2 f_1 \ll 1$, it follows that $r_1 \ll r_0$. Using the relation

$$K_0(ze^{im\pi}) = K_0(z) - im\pi I_0(z)$$

(Watson 1944, p. 80) we find from equation (3.3.66) that

$$\frac{\theta-\theta^{(0)}}{\theta^{(0)}}=\frac{i\varepsilon a\Omega}{\beta^2 f_1}\left[\frac{K_0(r_1)-K_0(r_0)-1/2\,I_0(r_0)-1/2i\pi\{I_0(r_1)-1/2\,I_0(r_0)\}}{K_0(r_0)-1/4iI_0(r_0)}\right]. \tag{3.3.68}$$

If we use the asymptotic expansions of the modified Bessel functions (Watson 1944, p. 202) we find that for large values of ϱ

$$\left|\frac{\theta-\theta^{(0)}}{\theta^{(0)}}\right|=\frac{a\Omega}{f_1\beta^2}\phi(r_0,r_1),$$

where

$$\phi(r_0,r_1)=\left\{\left[2\left(\frac{r_0}{r_1}\right)^{1/2}e^{r_1-r_0}-1\right]^2+\frac{4}{\pi^2}\right\}^{1/2}.$$

But $r_1 \ll r_0$ so that

$$\left|\frac{\theta-\theta^{(0)}}{\theta^{(0)}}\right|\simeq\frac{\varepsilon a\Omega}{f_1\beta^2}\left(1+\frac{4}{\pi^2}\right)^{1/2} \tag{3.3.69}$$

showing that for small values of $(a\Omega/f_1\beta^2)$ the quasi-static value of the temperature is a very good approximation to the exact value.

(ii) The Effect of a Moving Line Source

We shall consider now the effect of a line source of heat of constant strength which is moving with uniform velocity V in a direction perpendicular to its own length. If the source remains parallel to the x_3-axis and if its velocity is along the x_1-axis we may write, in the notation of equation (3.3.19)

$$\Theta = F\delta(x_2)\delta(x_1 - pt) ,$$

where

$$p = V\tau/l .$$

For this function

$$\hat{\Theta}(\xi_1, \xi_2, \omega) = \frac{F}{(2\pi)^{1/2}}\delta(\omega + p\xi_1)$$

so that from equation (3.3.20) we obtain the quasi-static solution

$$u_q^{(0)} = \frac{bF}{4\pi^2\beta^2}\int_{-\infty}^{\infty}\int_{-\infty}^{\infty}\frac{i\xi_q \exp[-i\{(x_1 - pt)\xi_1 + x_2\xi_2\}]}{(\xi_1^2 + \xi_2^2)(\xi_1^2 + \xi_2^2 + if_1 p\xi_1)} d\xi_1 d\xi_2 \ . \ (q = 1,2) \qquad (3.3.70)$$

For this solution the components of stress $\sigma_{11}^{(0)}$, $\sigma_{22}^{(0)}$, $\sigma_{12}^{(0)}$ are given by the equation

$$\sigma_{qr}^{(0)} = -\frac{Fb\delta_{qr}}{4\pi^2}\int_{-\infty}^{\infty}\int_{-\infty}^{\infty}\frac{\exp - i\{(x_1 - pt)\xi_1 + x_2\xi_2\}}{\xi_1^2 + \xi_2^2 + ifp\xi_1} d\xi_1 d\xi_2$$
$$+ \frac{bF}{4\pi^2\beta^2}\int_{-\infty}^{\infty}\int_{-\infty}^{\infty}\frac{[(\beta^2 - 2)(\xi_1^2 + \xi_2^2)\delta_{qr} + 2\xi_q\xi_r + ibgp\xi_1(\xi_1^2 + \xi_2^2)(\xi_1^2 + \xi_2^2 + ifp\xi_1)^{-1}]\delta_{pq}}{(\xi_1^2 + \xi_2^2)(\xi_1^2 + \xi_2^2 + if_1 p\xi_1)}$$
$$\times \exp[-i\{(x_1 - pt)\xi_1 + x_2\xi_2\}] d\xi_1 d\xi_2 \ . \qquad (3.3.71)$$

For example, we find that

$$\sigma_{12}^{(0)} = \frac{bF}{2\pi^2\beta^2}\int_{-\infty}^{\infty}\int_{-\infty}^{\infty}\frac{\xi_1\xi_2\exp[-i\{(x_1-pt)\xi_1+x_2\xi_2\}]d\xi_1 d\xi_2}{(\xi_1^2+\xi_2^2)(\xi_1^2+\xi_2^2+if_1p\xi_1)} . \tag{3.3.72}$$

Performing the integration with respect to ξ_1 we find that

$$\int_{-\infty}^{\infty}\int_{-\infty}^{\infty}\frac{\xi_1\xi_2\exp[-i\{(x_1-pt)\xi_1+x_2\xi_2\}]d\xi_1 d\xi_2}{(\xi_1^2+\xi_2^2)(\xi_1^2+\xi_2^2+if_1p\xi_1)} =$$

$$= \frac{2\pi}{f_1 p}\int_0^{\infty}\left\{\frac{2\xi_2}{(4\xi_2^2+p^2)^{1/2}}e^{-1/2(x_1-pt)[(4\xi_2^2+f_1^2p^2)^{1/2}+f_1p]}-e^{-\xi_2(x_1-pt)}\right\}\sin(\xi_2 x_2)d\xi_2 ,$$

and it can be shown that this integral has the value

$$\frac{2\pi x_2}{f_1 pR}\left[\frac{2}{f_1 p}e^{-1/2\, f_1p(x_1-pt)}K_1\left(\frac{1}{2}f_1pR\right)-1\right], \tag{3.3.73}$$

where R is defined by the equation

$$R^2 = (x_1-pt)^2+x_2^2 . \tag{3.3.74}$$

Substituting from equation (3.3.72) we find that

$$\sigma_{12}^{(0)} = \frac{bx_2F}{\pi\beta^2 f_1pR}\left[\frac{2}{f_1p}e^{-1/2\, f_1p(x_1-pt)}K_1\left(\frac{1}{2}f_1pR\right)-1\right] . \tag{3.3.75}$$

In a similar way we can establish the result

$$\int_{-\infty}^{\infty}\int_{-\infty}^{\infty}\frac{e^{-i(x_1-pt)\xi_1-ix_2\xi_2}}{\xi_1^2+\xi_2^2+ipf\xi_1}d\xi_1 d\xi_2 = 4\pi\int_0^{\infty}\frac{e^{-1/2(x_1-pt)[(4\xi_2^2+f^2p^2)^{1/2}+fp]}}{(4\xi_2^2+f^2p^2)^{1/2}}d\xi_2 =$$

$$= \frac{2\pi(x_1-pt)}{R} K_1\left(\frac{1}{2} fpR\right) e^{-1/2\, fp(x_1-pt)} \tag{3.3.76}$$

from which we obtain the equations

$$\sigma_{11}^{(0)} + \sigma_{22}^{(0)} = \frac{bF}{\pi}\left[\frac{\beta^2-1}{\beta^2} e^{-1/2\, f_1p(x_1-pt)} K_0\left(\frac{1}{2} f_1 pR\right) - e^{-1/2\, fp(x_1-pt)} K_0\left(\frac{1}{2} fpR\right)\right]$$
$$+ \frac{b^2 gF}{2\pi\beta^2}\left(\frac{x_1-pt}{R}\right)\left\{K_1\left(\frac{1}{2} fpR\right) e^{-1/2\, fp(x_1-pt)} - K_1\left(\frac{1}{2} f_1 pR\right) e^{-1/2\, f_1p(x_1-pt)}\right\}, \tag{3.3.77}$$

$$\sigma_{11}^{(0)} - \sigma_{22}^{(0)} = \frac{2bF}{\pi\beta^2 f_1 pR}\left[\frac{2}{f_1 p} e^{-1/2\, f_1p(x_1-pt)} K_1\left(\frac{1}{2} f_1 pR\right) - \frac{1}{R}\right] \tag{3.3.78}$$

by means of which we may calculate the stress components $\sigma_{11}^{(0)}$ and $\sigma_{22}^{(0)}$.

(iii) The Stress due to an Impulsive Line Source.

We shall now consider the effect of an impulsive line source of strength F applied along the line $x_1 = x_2 = 0$. This may be represented by

$$\Theta(x_1, x_2, t) = F\delta(x_1)\delta(x_2)\delta(t) \tag{3.3.79}$$

from which, in the notation of equation (3.3.19)

$$\hat{\Theta} = \frac{F}{(2\pi)^{3/2}}.$$

Substituting from equation (3.3.79) into equation (3.3.20) we obtain the solution

$$u_p = \frac{bF}{8\pi^3}\int_{T_3}\frac{i\xi_{\hat{p}}\exp-i(\xi_1x_1+\xi_2x_2+\omega t)d\vec{T}}{\beta^2(\xi_1^2+\xi_2^2)^2-(i\omega f_1\beta^2+a\omega^2)(\xi_1^2+\xi_2^2)+iaf\omega^3} . \qquad (p=1,2)$$

Alternatively if we use cylindrical co-ordinates then

$$\Theta = \frac{F}{2\pi\varrho}\delta(\varrho)\delta(t)$$

so that

$$\hat{\Theta} = \frac{F\delta(\zeta)}{2\pi} .$$

Inserting this expression in equations (3.3.49), (3.3.50) and (3.3.51) we find that $u_z = 0$ and that u_ϱ and Θ are given by the equations

$$u_\varrho = \frac{bF}{4\pi^2}\int_{-\infty}^{\infty}e^{-i\omega t}d\omega\int^{\infty}\frac{\xi^2J_1(\xi\varrho)d\xi}{\mathfrak{D}(\xi,\omega)} , \qquad (3.3.80)$$

$$\Theta = \frac{F}{4\pi^2}\int_{-\infty}^{\infty}e^{-i\omega t}d\omega\int_0^{\infty}\frac{(\beta^2\xi^2-a\omega^2)\xi J_0(\xi\varrho)d\xi}{\mathfrak{D}(\xi,\omega)} , \qquad (3.3.81)$$

where

$$\mathfrak{D} = (\beta^2\xi^2-a\omega^2)(\xi^2-if\omega)-i\omega bg\xi^2 . \qquad (3.3.82)$$

If we assume that $a=0$ we find that equation (3.3.80) reduces to

$$u_\varrho^{(0)} = \frac{bF}{2\pi\beta^2 f_1}\int_0^\infty e^{-\xi^2 t/f_1} J_1(\xi\varrho)\,d\xi \ .$$

Now

$$\int_0^\infty J_1(\lambda\varrho)e^{-k\lambda^2}\,d\lambda = \varrho^{-1}\{1-\exp(-\varrho^2/4k)\}H(k)$$

so that $u_\varrho^{(0)} = 0$ if $t<0$ and

$$u_\varrho^{(0)} = \frac{bF}{2\pi\beta^2 f_1\varrho}\{1-\exp(-f_1\varrho^2/4t)\}\ , \quad t>0 \ .$$

Similarly, from equation (3.3.81) we obtain the approximate formula

$$\theta^{(0)} = (F/4\pi t)\exp(-f_1\varrho^2/4t)H(t) \ .$$

Since we have assumed that $a=0$ we should expect these solutions to be valid if $\varrho \ll V_s t$. For very short times, i.e. immediately after the application of the thermal shock these expressions would not be valid.

The corresponding components of stress for $t \gg \varrho/V_s$ are

$$\sigma_{\varrho\varrho}^{(0)} = -\frac{bF}{\pi\beta^2 f_1\varrho^2}\{1-\exp(-f_1\varrho^2/4t)\}$$

$$\sigma_{\phi\phi}^{(0)} = -\frac{bF}{2\pi\beta^2 f_1 \varrho^2}\frac{f_1}{t}\exp(-f_1\varrho^2/4t)-2\{1-\exp(-f_1\varrho^2/4t)\}$$

$$\sigma_{zz}^{(0)} = -\frac{bF}{2\pi\beta^2 t}\exp(-f_1\varrho^2/4t)$$

and the shearing stress $\sigma_{\varrho z}$ is identically zero.

(iv) Impulsive Point Source of Heat.

To illustrate further the use of the axially symmetrical solution we shall consider the effect of an impulsive source of strength q situated at the origin. For such a source we have

$$\Theta = \frac{q}{2\pi\varrho}\delta(\varrho)\delta(z)\delta(t)$$

so that,

$$\hat{\Theta} = \frac{q}{2\pi}\,. \tag{3.3.83}$$

Substituting from equation (3.3.83) into equation (3.3.49) we obtain the approximate solution, valid for all but very large values of r/t ,

$$u_\varrho^{(0)} = \frac{bq}{4\pi\beta^2}\int_{-\infty}^{\infty}\int_{-\infty}^{\infty} e^{-i(\zeta z+\omega t)}d\zeta d\omega \int_0^{\infty}\frac{\xi^2 J_1(\xi\varrho)d\xi}{(\xi^2+\zeta^2)(\xi^2+\zeta^2-if_1\omega)}\,.$$

Performing the integration with respect to ω we find that

$$u^{(0)} = \frac{bq}{\pi\beta^2 f_1}\int_0^\infty \xi^2 J_1(\xi r)e^{-\xi^2 t/f_1}d\xi\int_0^\infty \frac{\cos(z)e^{-\zeta^2 t/f_1}d\zeta}{\xi^2+\zeta^2}$$

and then performing the integration with respect to ζ (Erdelyi *et al.* 1954, p. 15), we obtain for the radial component of the displacement the expression

$$u_\varrho^{(0)} = \frac{bq}{4\beta^2 f_1}\int_0^\infty \xi J_1(\xi r)\left[e^{-\xi z}\mathrm{Erfc}\left\{\frac{t^{1/2}\xi}{f_1^{1/2}}-\frac{zf_1^{1/2}}{2t^{1/2}}\right\}+e^{\xi z}\mathrm{Erfc}\left\{\frac{t^{1/2}\xi}{f_1^{1/2}}+\frac{zf^{1/2}}{2t^{1/2}}\right\}\right]d\xi$$

from which numerical values may be obtained by quadratures.

Similarly it can be shown that

$$u_z^{(0)} = \frac{bq}{4\beta^2 f_1}\int_0^\infty \xi J_0(\xi r)\left[e^{-\xi z}\mathrm{Erfc}\left\{\frac{t^{1/2}\xi}{f_1^{1/2}}-\frac{zf_1^{1/2}}{2t^{1/2}}\right\}-e^{\xi z}\mathrm{Erfc}\left\{\frac{t^{1/2}\xi}{f_1^{1/2}}+\frac{zf_1^{1/2}}{2t^{1/2}}\right\}d\xi\right. .$$

3.4 The Stresses Produced in an Infinite Elastic Solid by Body Forces

In this section we consider the generation of thermoelastic disturbances by the action of body forces. (Cf. Lockett & Sneddon, 1959).

If we eliminate the stresses between (3.3.1) and (3.3.2) we get an equation

$$(\beta^2-2)\vartheta_{,p}-b\Theta_{,p}+2\varepsilon_{pq,q}+F_p = a\ddot{u}_p \qquad (3.4.1)$$

which can be differentiated with respect to x_r to give on contracting,

$$\beta^2 \Delta_3 \vartheta - b\Delta_3 \Theta + F_{p,p} = a\ddot{\vartheta} . \tag{3.4.2}$$

Defining multiple Fourier transforms as before we find that the equations (3.4.1), (3.4.2), (3.3.3) are equivalent to the algebraic equations

$$-i\xi_p(\beta^2-2)\hat{\vartheta} + ib\xi_p\hat{\Theta} - \xi^2\hat{u}_p - i\xi_p\hat{\vartheta} + \hat{F}_p = -a\omega^2\hat{u}_p \tag{3.4.3}$$

$$-\beta^2\xi^2\hat{\vartheta} + b\xi^2\hat{\Theta} - i\xi_p\hat{F}_p = -a\omega^2\hat{\vartheta} \tag{3.4.4}$$

$$-\xi^2\hat{\Theta} = -i\omega f\hat{\Theta} - i\omega g\hat{\vartheta} \tag{3.4.5}$$

where we have assumed that $\Theta = 0$ and written $\xi^2 = \xi_p\xi_p$. The solution of the pair of equations (3.4.3), (3.4.5) is given by the equations

$$D(\omega,\xi^2)\hat{\vartheta} = -i(\xi^2 - i\omega f)\xi_q\hat{F}_q \tag{3.4.6}$$

$$D(\omega,\xi^2)\hat{\Theta} = \omega g\xi_q\hat{F}_q , \tag{3.4.7}$$

in which

$$D(\omega,\xi^2) = (\beta^2\xi^2 - a\omega^2)(\xi^2 - i\omega f) - i\omega bg\xi^2 . \tag{3.4.8}$$

Substituting these forms into equation (3.4.4) we find that

$$\hat{u}_p = (\xi^2 - a\omega^2)^{-1}\hat{F}_p - \frac{D_1(\omega,\xi^2)}{D(\omega,\xi^2)}\xi_p\xi_q\hat{F}_q \tag{3.4.9}$$

where we have written

$$D_1(\omega,\xi^2) = (\beta^2-1)(\xi^2-i\omega f)-i\omega bg\,. \qquad (3.4.10)$$

Similarly from equations (3.3.27), (3.3.28), (3.3.29) we deduce that in the absence of heat sources but in the presence of a distribution of body forces described by the vector field $(F_\varrho,0,F_z)$, the transforms $\bar{u}_\varrho, \bar{u}_z$ defined by equations (3.3.30) and (3.3.31) satisfy the algebraic equations

$$(\beta^2\xi^2+\zeta^2-a\omega^2)\bar{u}_\varrho - i(\beta^2-1)\xi\zeta\bar{u}_z - \bar{F}_\varrho = b\xi\bar{\theta}$$

$$i(\beta^2-1)\xi\zeta\bar{u}_\varrho + (\xi^2+\beta^2\zeta^2-a\omega^2)\bar{u}_z - \bar{F}_z = ib\zeta\bar{\theta}\,.$$

Similarly if $\theta \equiv 0$, equation (3.3.27) is equivalent to the equation

$$(\xi^2+\zeta^2-i\omega f)\bar{\theta} = i\omega g(\xi\bar{u}_\varrho - i\zeta\bar{u}_z)\,.$$

The solution of these equations is

$$\bar{u}_\varrho = \frac{D_1(\omega,\xi^2,\zeta^2)}{D(\omega,\xi^2,\xi^2)}\bar{F}_\varrho + i\frac{D_2(\omega,\xi^2,\zeta^2)}{D(\omega,\xi^2,\zeta^2)}\bar{F}_z \qquad (3.4.11)$$

$$\bar{u}_z = -i\frac{D_1(\omega,\xi^2,\zeta^2)}{D(\omega,\xi^2,\zeta^2)}\bar{F}_\varrho + \frac{D_1(\omega,\zeta^2,\xi^2)}{D(\omega,\xi^2,\zeta^2)}\bar{F}_z \qquad (3.4.12)$$

$$\bar{\theta} = \frac{\omega g}{D(\omega,\xi^2,\zeta^2)}(i\xi\bar{F}_\varrho + \zeta\bar{F}_z) \qquad (3.4.13)$$

where D_1, D_2 and D are defined by the equations

$$(\xi^2+\zeta^2-a\omega^2)D_1(\omega,\xi^2,\zeta^2)=(\xi^2+\beta^2\zeta^2-a\omega^2)(\xi^2+\zeta^2-i\omega f)-i\omega bg\zeta^2$$

$$(\xi^2+\zeta^2-a\omega^2)D_2(\omega,\xi^2,\zeta^2)=\{(\beta^2-1)(\xi^2+\zeta^2-i\omega f)-i\omega bg\}\xi\zeta$$

$$D(\omega,\xi^2,\zeta^2)=(\beta^2\xi^2+\beta^2\zeta^2-a\omega^2)(\xi^2+\zeta^2-i\omega f)-i\omega bg(\xi^2+\zeta^2) .$$

The corresponding expressions for the displacement components u_ϱ, u_z and the temperature Θ are obtained by using the relevant inversion formulae.

For example, since in the *quasi-static approximation* $a = 0$ and

$$D=\beta^2(\xi^2+\zeta^2)(\xi^2+\zeta^2-i\omega f_1)$$

where $f_1 = f(1+\varepsilon)$, $\varepsilon = bg/\beta^2 f$, so that in the case in which $F_\varrho \equiv 0$ equation (3.4.13) reduces to

$$\bar{\Theta}^{(0)}=\frac{g\omega\zeta}{\beta^2(\xi^2+\zeta^2)(\xi^2+\zeta^2-i\omega f_1)}\bar{\bar{F}}_z .$$

Comparing this equation with equation (3.3.51) we see that the quasi-static solution for a body-force $(0,0,F_z)$ can be obtained from a solution of the classical equation for the conduction of heat

$$\Delta_2\,\Theta+\Theta=f\frac{\partial\Theta}{\partial t}$$

by replacing f by f_1 and taking for Θ the source function

$$\bar{\bar{\Theta}}_F = \frac{g\omega\zeta}{\beta^2(\xi^2+\zeta^2)}\bar{\bar{F}}_z .$$

For example if we have a point-force at the origin we may take

$$F_z = (2\pi\varrho)^{-1}\delta(\varrho)\delta(z)f(t)$$

and

$$\hat{F}_z = (2\pi)^{-3/2}\hat{f}(\omega)$$

where

$$\hat{f}(\omega) = \mathfrak{F}[f(t);\omega] .$$

Hence

$$\bar{\bar{\Theta}}_F = \frac{g}{\beta^2(2\pi)^{3/2}}\cdot\frac{\zeta}{(\xi^2+\zeta^2)}\,\omega\hat{f}(\omega)$$

so that applying the inversion theorems for Fourier and Hankel transforms we have the relation

$$\Theta_F = \frac{gf'(t)}{2\pi^2\beta^2}\int_0^\infty\int_0^\infty \frac{\xi\zeta}{\xi^2+\zeta^2}J_0(\xi\varrho)\sin(\zeta z)\,d\xi\,d\zeta$$
$$= \frac{gf'(t)}{4\pi\beta^2}z(\varrho^2+z^2)^{-3/2} .$$

This result can be immediately generalized to give, *in the quasi-static approximation*, the heat source - equivalent to the distribution body force $(0,0,F_z)$ with

$$F_z = \phi(x,y,z)f(t) .$$

This source is given by

$$\Theta = f'(t)\int_{E_3} \phi(\vec{x})G(\vec{x}-\vec{x}')d\vec{x}'$$

where the Green's function G is defined by the equation

$$G(\vec{x}) = \frac{g}{4\pi\beta^2} zr^{-3} , \quad r = |\vec{x}| .$$

3.5 The Stresses Produced in an Elastic Half-Space by Uneven Heating

We shall now consider the case in which the solid is bounded by a plane which is free from applied stress but whose surface temperature is made to vary in a prescribed way and which has heat sources of known strength distributed throughout the interior of the half-space.

In the case of a semi-infinite solid the symmetry of the equations, when written in cartesian coordinates, is not preserved, so that it is no longer advantageous to make use of the notation x_i ,$(i = 1, 2, 3)$ employed above. In this section we shall denote the coordinates of a typical point of the solid by (x,y,z) and assume that the solid is bounded by the plane $z = 0$

and occupies the space $z \geqslant 0$. If we denote the components of the displacement vector by (u_x, u_y, u_z) and those of the stress tensor by $\sigma_{xx}, \sigma_{yy}, \sigma_{zz}, \sigma_{yz}, \sigma_{xz}, \sigma_{xy}$ we may write the equations of motion (3.3.1) in the forms

$$\frac{\partial \sigma_{xx}}{\partial x} + \frac{\partial \sigma_{xy}}{\partial y} + \frac{\partial \sigma_{xz}}{\partial z} = a\frac{\partial^2 u_x}{\partial t^2} \qquad (3.5.1)$$

$$\frac{\partial \sigma_{xy}}{\partial x} + \frac{\partial \sigma_{yy}}{\partial y} + \frac{\partial \sigma_{yz}}{\partial z} = a\frac{\partial^2 u_y}{\partial t^2} \qquad (3.5.2)$$

$$\frac{\partial \sigma_{xz}}{\partial x} + \frac{\partial \sigma_{yz}}{\partial z} + \frac{\partial \sigma_{zz}}{\partial z} = a\frac{\partial^2 u_z}{\partial t^2} \qquad (3.5.3)$$

(in the absence of body forces). The stress-strain relations take the form

$$\sigma_{xx} = \beta^2\frac{\partial u_x}{\partial x} + (\beta^2 - 2)\left[\frac{\partial u_y}{\partial y} + \frac{\partial u_z}{\partial z}\right] - b\theta , \qquad (3.5.4)$$

$$\sigma_{yy} = (\beta^2 - 2)\frac{\partial u_x}{\partial x} + \beta^2\frac{\partial u_y}{\partial y} + (\beta^2 - 2)\frac{\partial u_z}{\partial z} - b\theta , \qquad (3.5.5)$$

$$\sigma_{zz} = (\beta^2 - 2)\left[\frac{\partial u_z}{\partial z} + \frac{\partial u_y}{\partial y}\right] + \beta^2\frac{\partial u_z}{\partial z} - b\theta , \qquad (3.5.6)$$

$$\sigma_{yz} = \frac{\partial u_z}{\partial y} + \frac{\partial u_y}{\partial z} , \qquad (3.5.7)$$

$$\sigma_{xz} = \frac{\partial u_x}{\partial z} + \frac{\partial u_z}{\partial x} , \tag{3.5.8}$$

$$\sigma_{xy} = \frac{\partial u_y}{\partial x} + \frac{\partial u_x}{\partial y} , \tag{3.5.9}$$

where the temperature variation Θ satisfies the equation

$$\Delta_3\Theta + \Theta = f\frac{\partial\Theta}{\partial t} + g\frac{\partial}{\partial t}\left(\frac{\partial u_x}{\partial x} + \frac{\partial u_y}{\partial y} + \frac{\partial u_z}{\partial z}\right) . \tag{3.5.10}$$

If we introduce the sets of transforms

$$(\hat\sigma_{xx}, \hat\sigma_{yy}, \hat\sigma_{zz}, \hat\sigma_{xy}, \hat u_x, \hat u_y, \hat\Theta) = \tag{3.5.11}$$
$$= \mathfrak{F}_{(3)}\left[\mathfrak{F}_s\{\sigma_{xx}, \sigma_{yy}, \sigma_{zz}, \sigma_{xy}, u_x, u_y, \Theta, z \to \zeta\}; (x,y,t) \to (\xi,\eta,\omega)\right]$$

$$(\hat\sigma_{xz}, \hat\sigma_{yz}, \hat u_z) = \mathfrak{F}_{(3)}\left[\mathfrak{F}_c\{\sigma_{x}, \sigma_{yz}, u_z; z \to \zeta\}; (x,y,t) \to (\xi,\eta,\omega)\right] \tag{3.5.12}$$

then multiplying equations (3.5.1) and (3.5.2) by

$$\exp[i(\xi x + \eta y + \omega t)]\sin(\zeta z)$$

and equation (3.5.3) by

$$\exp[i(\xi x + \eta y + \omega t)]\cos(\zeta z)$$

and integrating over all t throughout the half-space and assuming that

$$\sigma_{zz}(x,y,0,t) = 0 \tag{3.5.13}$$

we find that

$$i\xi\hat{\sigma}_{xx} + i\eta\hat{\sigma}_{xy} + \zeta\hat{\sigma}_{xz} = a\omega^2\hat{u}_x$$

$$i\xi\hat{\sigma}_{xy} + i\eta\hat{\sigma}_{yy} + \zeta\hat{\sigma}_{yz} = a\omega^2\hat{u}_y \tag{3.5.14}$$

$$i\xi\sigma_{xz} + i\eta\hat{\sigma}_{yz} + \zeta\hat{\sigma}_{zz} = a\omega^2\hat{u}_z .$$

Similarly if we assume that

$$u_x(x,y,0,t) = u_y(x,y,0,t) = 0 \tag{3.4.15}$$

we find that the partial differential equations (3.5.4) to (3.5.9) are equivalent to the following set of six algebraic equations

$$\hat{\sigma}_{xx} = -i\beta^2\xi\hat{u}_x - i(\beta^2-2)(\eta\hat{u}_y - i\zeta\hat{u}_z) - b\hat{\theta}$$

$$\hat{\sigma}_{yy} = -i(\beta^2-2)\xi\hat{u}_x - i\beta^2\eta\hat{u}_y - (\beta^2-2)\zeta\hat{u}_z - b\hat{\theta}$$

$$\hat{\sigma}_{zz} = -i(\beta^2-2)(\xi\hat{u}_x + \eta\hat{u}_y) - \beta^2\zeta\hat{u}_z - b\hat{\theta} \tag{3.5.16}$$

$$\hat{\sigma}_{yz} = (\zeta\hat{u}_y - i\eta\hat{u}_z)$$

$$\hat{\sigma}_{xz} = (-i\xi\hat{u}_z + \zeta\hat{u}_x)$$

$$\hat{\sigma}_{xy} = -i(\eta\hat{u}_x + \xi\hat{u}_y) .$$

With the same assumptions about the surface values of u_x and u_y we find on multiplying both sides of (3.5.10) by

$$\exp\{i(\xi x + \eta y + \omega t)\}\sin(\zeta z)$$

and integrating that

$$\hat{\dot{\Theta}} - (\xi^2 + \eta^2 + \zeta^2)\hat{\Theta} + \zeta\hat{\Theta}_0 = -if\omega\hat{\Theta} - g\xi\omega\hat{u}_x - g\eta\omega\hat{u}_y + ig\omega\zeta\hat{u}_z$$

where (3.5.17)

$$\hat{\Theta} = \mathfrak{F}_{(3)}\left[\mathfrak{F}_s\{\Theta(x,y,z,t); z \to \zeta\}; (x,y,t) \to (\xi,\eta,\omega)\right]$$

$$\hat{\Theta}_0 = \mathfrak{F}_{(3)}\left[\Theta(x,y,0,t); (x,y,t) \to (\xi,\eta,\omega)\right] \tag{3.5.18}$$

in which $\Theta(x\ y\ 0\ t)$ denotes the surface temperature.

Solving the set of algebraic equations (3.5.14), (3.5.16) and (3.5.17) we find that the transforms of the components of the displacement vector are given by the equations

$$(\hat{u}_x, \hat{u}_y, \hat{u}_z) = (i\xi, i\eta, -\zeta) b\hat{G} \tag{3.5.19}$$

where

$$\hat{G} = \frac{\hat{\dot{\Theta}} + \zeta\hat{\Theta}_0}{(\gamma^2 - if\omega)(\beta^2\gamma^2 - a\omega^2) - ibg\omega\gamma^2} \tag{3.5.20}$$

with $\gamma^2 = \xi^2 + \eta^2 + \zeta^2$. Inverting equations (3.5.19) by means of Fourier's inversion theorem for multiple transforms we find that the components of the displacement vector are given by the equations

$$u_x = (2\pi^2)^{-1} b \int_H i\xi \exp\{-i(\xi x + \eta y + \omega t)\} \sin(\zeta z)\hat{G}\,dh \tag{3.5.21}$$

$$u_y = (2\pi^2)^{-1} b \int_H i\eta \exp\{-i(\xi x + \eta y + \omega t)\} \sin(\zeta z)\hat{G}\,dh \tag{3.5.22}$$

$$u_z = (2\pi^2)^{-1} b \int_H \zeta \exp\{-i(\xi x + \eta y + \omega t)\} \cos(\zeta z) \hat{G} dh \qquad (3.5.23)$$

where

$$H = \{(\xi,\eta,\omega) : (\xi,\eta,\omega) \in R^3, \quad \zeta \in R^+\}$$

and $dh = d\xi d\eta d\zeta d\omega$.

This set of equations can be written in the vector form

$$\vec{u} = -\text{grad}\, \Psi \qquad (3.5.24)$$

with

$$\Psi(x,y,z,t) = (2\pi^2)^{-1} b \int_H \exp\{-i(\xi x + \eta y + \omega t)\} \sin(\zeta z) \hat{G} dh \, . \qquad (3.5.25)$$

The equations (3.5.24) and (3.5.25) constitute the solution of the problem in which there is a prescribed source function Θ and a prescribed surface distribution of temperature $\theta(x,y,0,t)$ and the mechanical boundary conditions are defined by equation (3.5.13) and (3.5.15). For this solution we find that

$$\sigma_{xz}(x,y,0,t) = \pi^{-2} b \int_H i \xi \zeta \hat{G} \exp\{-i(\xi x + \eta y + \omega t)\} dh \qquad (3.5.26)$$

$$\sigma_{yz}(x,y,0,t) = \pi^{-2} b \int_H i \eta \zeta G \exp\{-i(\xi x + \eta y + \omega t)\} dh \, .$$

If, therefore, we wish to find a solution to the problem in which

$$\sigma_{xz} = \sigma_{yz} = \sigma_{zz} = 0 \quad \text{on} \quad z = 0 \qquad (3.5.27)$$

we have to add to the solution (3.5.24) the solution of the purely elastic boundary value problem in which

$$\sigma_{xz} = -q_1(x,y,t)\,, \quad \sigma_{yz} = -q_2(x,y,t)\,, \quad \sigma_{zz} = 0 \text{ on } z = 0 \tag{3.5.28}$$

where q_1 and q_2 are defined by equations (3.5.26) and (3.5.27) respectively. The solution of the half-space problem corresponding to the boundary conditions (3.5.28) has been derived by Eason (1954). It should be noted that for the functions involved in this particular form of the problem

$$\mathfrak{F}_{(3)}[(q_1,q_2):(x,y,t)\rightarrow(\xi,\eta,\omega)] = (\xi,\eta)I(\xi,\eta,\omega) \tag{3.5.29}$$

where

$$I(\xi,\eta,\omega) = ib(2/\pi)^{1/2}\int_0^\infty \frac{\zeta(\hat{\Theta}+\Theta_0)d\zeta}{(\gamma^2-if\omega)(\beta^2\gamma^2-a\omega^2)-ibg\omega\gamma^2}\,. \tag{3.5.30}$$

We may obtain the solution of the two-dimensional thermoelastic equations for the half-plane $\{(x,z): z>0\}$ by substituting in equations (3.5.21) and (3.5.23) the expressions

$$\hat{\Theta} = (2\pi)^{1/2}\bar{q}(\xi,\zeta,\omega)\delta(\eta) \qquad \hat{\Theta}_0 = \bar{\Theta}_0(\xi,\omega)\delta(\eta)$$

where

$$\bar{q}(\xi,\zeta,\omega) = \mathfrak{F}_{(2)}[\mathfrak{F}_s\{\Theta(x,z,t);z\rightarrow\zeta\};(x,t)\rightarrow(\xi,\omega)] \tag{3.5.31}$$

and

$$\bar{\Theta}_0(\xi,\omega) = \mathcal{F}_{(2)}\left[\Theta(x,0,t):(x,t)\rightarrow(\xi,\omega)\right]. \qquad (3.5.32)$$

In this way we obtain the solution

$$u_x = (2\pi^2)^{-1} b \int_{-\infty}^{\infty}\int_{-\infty}^{\infty} i\xi \exp\{-i(\xi x+\omega t)\} d\xi d\omega \int_0^{\infty} \bar{F}\sin(\zeta z) d\zeta , \qquad (3.5.33)$$

$$u_z = -(2\pi^2)^{-1} b \int_{-\infty}^{\infty}\int_{-\infty}^{\infty} \exp\{-i(\xi x+\omega t)\} d\xi d\omega \int_0^{\infty} \zeta\bar{F}\cos(\zeta z) d\zeta , \qquad (3.5.34)$$

where $\bar{F}$ is defined by the equation

$$\{(\xi^2+\zeta^2-if\omega)(\beta^2\xi^2+\beta^2\zeta^2-a\omega^2)-ibg\omega(\xi^2+\zeta^2)\}\bar{F}(\xi,\zeta,\omega) = (2\pi)^{1/2}\bar{q}+\zeta\bar{\Theta}_0 . \qquad (3.5.35)$$

It is immediately obvious that this solution can be written in the form

$$u_x = -\frac{\partial\Psi}{\partial x}, \quad u_z = -\frac{\partial\Psi}{\partial z}. \qquad (3.5.36)$$

with

$$\Psi(x,z) = (2\pi^2)^{-1} b \int_{-\infty}^{\infty}\int_{-\infty}^{\infty} \exp\{-i(\xi x+\omega t)\} d\xi d\omega \int_0^{\infty} \zeta\bar{F}\sin(\zeta z) d\zeta . \qquad (3.5.37)$$

For this solution we have on $z=0$

$$\sigma_{xz} = (2\pi^2)^{-1} b \int_{-\infty}^{\infty}\int_{-\infty}^{\infty} i\xi\exp\{-i(\xi x+\omega t)\} d\xi d\omega \int_0^{\infty} \zeta\bar{F} d\zeta ,$$

$$\sigma_{zz} = 0 ,$$

so that to obtain the solution corresponding to a boundary free from applied stress we must add to the solution (3.5.36) that solution of the (purely elastic) dynamical equations which gives

$$\sigma_{xz}(x,0,t) = -2\tau(x,t)\,, \quad \sigma_{zz}(x,0,t) = 0 \qquad (3.5.38)$$

where

$$\bar{\tau} = \mathfrak{F}_{(2)}\left[\frac{1}{2}\sigma_{xz}(x,0,t);(x,t)\rightarrow(\xi,\omega)\right] = (bi\xi/\pi)\int_0^\infty \zeta \bar{F} d\zeta \,. \qquad (3.5.39)$$

The solution to this latter problem is (cf. Eason, 1954)

$$u_x(x,z,t) = (\beta^2-1)^{-1}\mathfrak{F}_{(2)}[|\xi|^{-1}\bar{\tau}\exp(-|\xi|z)\{\beta^2-(\beta^2-1)|\xi|z\};(\xi,\omega)\rightarrow(x,t)]$$

$$(3.5.40)$$

$$u_z(x,z,t) = (\beta^2-1)^{-1}\mathfrak{F}_{(2)}[i\xi^{-1}\bar{\tau}\exp(-|\xi|z)\{1+(\beta^2-1)|\xi|z\};(\xi,\omega)\rightarrow(x,t)]$$

so that the complete solution to our two-dimensional problem is contained in the equations (3.5.36) and (3.5.40)

3.6 The Thermoelastic Problem for an Infinite Medium with a Spherical Cavity

To illustrate the use of spherical polar coordinates (r,θ,ϕ) in the solution of the coupled equations of thermoelasticity, we consider the problem of determining the distribution of stress induced in an infinite elastic solid with a spherical cavity when the cavity face is subjected to a thermal as well as a mechanical constraint. We denote the radius of the cavity by a , so the elastic solid is specified by $r \geq a$.

If we assume that the thermoelastic disturbances

are radially symmetrical the various physical variables will be functions of r alone. Since the angular coordinates θ, ϕ do not figure in the subsequent analysis we may continue to write θ for the temperature deviation $T-T_0$. We define a displacement vector in terms of a scalar function $\Psi(r,t)$ through the equations

$$u_r = \frac{\partial}{\partial r}\left(\frac{\Psi}{r}\right), \quad u_\theta \equiv 0, \quad u_\phi \equiv 0. \tag{3.6.1}$$

and the temperature deviation in terms of a second scalar function $\chi(r,t)$ by the equation

$$\theta = r^{-1}\chi. \tag{3.6.2}$$

The thermoelastic equations then take the form

$$V_T^2\frac{\partial^2\Psi}{\partial r^2} - \frac{\partial^2\Psi}{\partial t^2} = (\alpha/\sigma\varkappa)\chi, \tag{3.6.3}$$

$$k\frac{\partial^2\chi}{\partial r^2} - \sigma c_\varepsilon\frac{\partial\chi}{\partial t} = (\alpha T_0/\varkappa)\frac{\partial^3\Psi}{\partial t\,\partial r^2}. \tag{3.6.4}$$

We assume the initial conditions

$$u_r(r,0)\equiv 0, \quad \dot{u}_r(r,0)\equiv 0, \quad \theta(r,0)\equiv 0, \quad (r\geq a), \tag{3.6.5}$$

and the boundary conditions

$$\sigma_{rr}(a,t) = -s(t), \quad \theta(a,t) = f(t), \quad (t\geq 0). \tag{3.6.6}$$

The equations (3.6.5) are easily shown to be equivalent to

$$\chi(r,0) \equiv 0\,, \quad \Psi(r,0) \equiv 0\,, \quad \Psi_t(r,0) \equiv 0\,, \quad (r \geqslant a), \tag{3.6.7}$$

so that if we introduce the Laplace transforms

$$\bar{\chi}(r,p) = \mathcal{L}[\chi(r,t);t \rightarrow p], \quad \bar{\Psi}(r,p) = \mathcal{L}[\Psi(r,t);t \rightarrow p]$$

and make use of (A.26) together with these initial conditions we find that the pair of partial differential equations (3.6.3), (3.6.4) is equivalent to the pair of ordinary differential equations

$$(V_T^2 D_r^2 - p^2)\bar{\Psi} = (\alpha/\sigma\varkappa)\bar{\chi}$$
$$(kD_r^2 - \sigma c_\varepsilon p)\bar{\chi} = (\alpha T_0/\varkappa)pD_r^2\bar{\Psi}$$

in which

$$D_r = \frac{d}{dr}\,.$$

Similarly the Laplace transforms of the equations (3.6.6) can be shown to be the equivalent to

$$\sigma V_r^2 a^2 D_r \bar{\Psi}(a,p) - 4\sigma V_S^2(aD_r - 1)\bar{\Psi}(a,p) - (a^2\alpha/\varkappa)\bar{\chi}(a,p) = -\bar{s}(p)$$
$$\bar{\chi}(a,p) = a\bar{f}(p)\,.$$

If we impose the conditions at infinity

$$u_r(r,t) \rightarrow 0\,, \quad \theta(r,t) \rightarrow 0 \quad \text{as} \quad r \rightarrow \infty$$

we see that the appropriate solution of this pair of ordinary differential equations is

$$\bar{\Psi}(r,p) = A\exp\{-\zeta_1(r-a)\} + B\exp\{-\zeta_2(r-a)\}$$

$$\bar{\chi}(r,p) = \alpha^{-1}\sigma\varkappa V_T^2[(\zeta_1^2 - p^2/V_T^2)A\exp\{-\zeta_1(r-a)\} + (\zeta_2^2 - p^2/V_T^2)B\exp\{-\zeta_2(r-a)\}]$$

where ζ_1 and ζ_2 are positive square roots of the square roots of the equation

$$\zeta^4 - \{p^2/V_T^2 + (p\sigma c_\varepsilon/k)(1+\varepsilon)\}\zeta^2 + p^3\sigma c_\varepsilon/kV_T^2 = 0$$

and A and B are given by the formula

$$\mathfrak{D}A = (a^3/\sigma)(\zeta_2^2 - p^2/V_T^2)\bar{s}(p) + (a\alpha/\sigma\varkappa V_T^2)\{p^2a^2 + 4(a\zeta_2+1)V_S^2\}\bar{f}(p)$$

$$\mathfrak{D}B = -(a^3/\sigma)(\zeta_1^2 - p^2/V_T^2)\bar{s}(p) - (a\alpha/\sigma\varkappa V_T^2)\{p^2a^2 + 4(a\zeta_1+1)V_S^2\}\bar{f}(p)$$

in which

$$\mathfrak{D} = (\zeta_1 - \zeta_2)\{(\zeta_1+\zeta_2)(p^2a^2 + 4V_S^2) + 4(\zeta_1\zeta_2 + p^2/V_T^2)a\,v_S^2\}\ .$$

The calculation of the inverse Laplace transforms is very complicated and it seems to be impossible to evaluate them in terms of known functions even for the simplest forms of the function $s(t)$ and $f(t)$. Recourse has to be made to approximate methods of solution. Two such techniques are available:

(a) Perturbation expansion (See Lessen, 1956)

(b) Asymptotic expansion for small times (See Chadwick, 1960).

CHAPTER 4

STATIC PROBLEMS OF THERMOELASTICITY

4.1 Solution of the Static Equations in Terms of Harmonic Functions

If we assume that the displacement field and the temperature field are both time-independent the equations (3.3.1)-- (3.3.3) reduce to

$$\sigma_{pq,q} + F_p = 0 \,, \tag{4.1.1}$$

$$\sigma_{pq} = [(\beta^2 - 2)\vartheta - b\theta]\delta_{pq} + 2\varepsilon_{pq} \,, \tag{4.1.2}$$

$$\Delta_3 \theta + \Theta = 0 \tag{4.1.3}$$

where

$$\beta^2 = \frac{2(1-\eta)}{1-2\eta} \,, \qquad b = (3\beta^2 - 4)\alpha T_0 \,. \tag{4.1.4}$$

It follows immediately that if the solution of a coupled problem is known the solution of the corresponding static problem is obtained by taking $a = f = g = 0$ in the general solution.

Following Sneddon (1962) we introduce three harmonic functions χ , Φ and Ψ so that

$$\Delta_3 \chi = 0 \,, \quad \Delta_3 \Phi = 0 \,, \quad \Delta_3 \Psi = 0 \tag{4.1.5}$$

and express the components of the displacement vector in terms of them through the equations

$$u_x = \frac{\partial \chi}{\partial x} + \frac{\partial \phi}{\partial x} + (\beta^2 - 1) z \frac{\partial^2 \phi}{\partial x \partial z} + z \frac{\partial \Psi}{\partial x} ,$$

$$u_y = \frac{\partial \chi}{\partial y} + \frac{\partial \phi}{\partial y} + (\beta^2 - 1) z \frac{\partial^2 \phi}{\partial y \partial z} + z \frac{\partial \Psi}{\partial y} , \tag{4.1.6}$$

$$u_z = \frac{\partial \chi}{\partial z} - \beta^2 \frac{\partial \phi}{\partial z} + (\beta^2 - 1) z \frac{\partial^2 \phi}{\partial z^2} + z \frac{\partial \Psi}{\partial z} - \Psi ,$$

and the temperature field through the equation

$$\Theta = \frac{2}{b} \frac{\partial \Psi}{\partial z} . \tag{4.1.7}$$

It is easily verified that, for this form of the displacement and temperature field, the dilatation is given by the equation

$$\vartheta = -2 \frac{\partial^2 \phi}{\partial z^2} , \tag{4.1.8}$$

and the stress field by the equations

$$\sigma_{xx} = -2(\beta^2 - 2) \frac{\partial^2 \phi}{\partial z^2} + 2 \frac{\partial^2 \chi}{\partial x^2} + 2 \frac{\partial^2 \phi}{\partial x^2} + 2(\beta^2 - 1) z \frac{\partial^3 \phi}{\partial x^2 \partial z} + 2z \frac{\partial^2 \Psi}{\partial x^2} - 2 \frac{\partial \Psi}{\partial z} ,$$

(4.1.9)

$$\sigma_{yy} = -2(\beta^2-2)\frac{\partial^2\phi}{\partial z^2} + 2\frac{\partial^2\chi}{\partial y^2} + 2\frac{\partial^2\phi}{\partial y^2} + 2(\beta^2-1)z\frac{\partial^3\phi}{\partial y^2\partial z} + 2z\frac{\partial^2\Psi}{\partial y^2} - 2\frac{\partial\Psi}{\partial z},$$

$$\sigma_{zz} = -2(\beta^2-1)\frac{\partial^2\phi}{\partial z^2} + 2\frac{\partial^2\chi}{\partial z^2} + 2(\beta^2-1)z\frac{\partial^3\phi}{\partial z^3} + 2z\frac{\partial^2\Psi}{\partial y^2} - 2\frac{\partial\Psi}{\partial z};$$

$$\sigma_{yz} = 2(\beta^2-1)z\frac{\partial^3\phi}{\partial y\partial z^2} + 2z\frac{\partial^2\Psi}{\partial y\partial z} + 2\frac{\partial^2\chi}{\partial y\partial z},$$

$$\sigma_{xz} = 2(\beta^2-1)z\frac{\partial^3\phi}{\partial x\partial z^2} + 2z\frac{\partial^2\Psi}{\partial x\partial z} + 2\frac{\partial^2\chi}{\partial x\partial z}, \qquad (4.1.10)$$

$$\sigma_{xy} = 2\frac{\partial^2\phi}{\partial x\partial y} + 2(\beta^2-1)z\frac{\partial^3\phi}{\partial x\partial y\partial z} + 2z\frac{\partial^2\Psi}{\partial x\partial y} + 2\frac{\partial^2\chi}{\partial x\partial y}.$$

From these equations we see that this solution satisfies the boundary conditions

$$\sigma_{xz} = \sigma_{yz} = 0, \quad \text{on } z = 0, \qquad (4.1.11)$$

provided that we choose χ in such a way that $\partial\chi/\partial z = 0$ when $z=0$ Further, on $z=0$, we have

$$u_z = \left[\frac{\partial\chi}{\partial z} - \beta^2\frac{\partial\phi}{\partial z} - \Psi\right]_{z=0}, \qquad (4.1.12)$$

$$\sigma_{zz} = \left[2\frac{\partial^2 \chi}{\partial z^2} - (\beta^2-1)\frac{\partial^2 \phi}{\partial z^2} - \frac{\partial \Psi}{\partial z}\right]_{z=0} . \qquad (4.1.13)$$

If we take

$$\chi \equiv 0, \quad \Psi = -(\beta^2-1)\frac{\partial \phi}{\partial z}$$

we obtain the solution

$$u_x = \frac{\partial \phi}{\partial x}, \quad u_y = \frac{\partial \phi}{\partial y}, \quad u_z = -\frac{\partial \phi}{\partial z}, \quad \theta = -\frac{2(\beta^2-1)}{b}\frac{\partial^2 \phi}{\partial z^2} \qquad (4.1.14)$$

for which

$$\sigma_{xz} = \sigma_{yz} = \sigma_{zz} = 0$$

at all points of the body. This solution was first derived by Lur'e (1955). (See also Sneddon and Tait, 1961).

In problems in which there is axial symmetry, we may retain the third equation of the set (4.1.6) and replace the first two by the single equation

$$u_\rho = \frac{\partial \chi}{\partial \rho} + \frac{\partial \phi}{\partial \rho} + (\beta^2-1) z \frac{\partial^2 \phi}{\partial \rho \partial z} + z\frac{\partial \Psi}{\partial \rho} . \qquad (4.1.15)$$

For this form of the displacement field the component σ_{zz} of the axisymmetric stress tensor is given by the third equation of the set (4.1.9) and the remaining non-vanishing components are given by the equations

$$\sigma_{\rho\rho} = 2\frac{\partial^2\chi}{\partial\rho^2} - 2(\beta^2-1)\frac{\partial^2\phi}{\partial z^2} + 2\frac{\partial^2\phi}{\partial\rho^2} + 2(\beta^2-1)z\frac{\partial^3\phi}{\partial\rho^2\partial z} + 2z\frac{\partial^2\Psi}{\partial\rho^2} - 2\frac{\partial\Psi}{\partial z}, \tag{4.1.16}$$

$$\frac{1}{2}(\sigma_{\rho\rho}+\sigma_{\phi\phi}) = -\frac{\partial^2\chi}{\partial z^2} - (2\beta^2-3)\frac{\partial^2\phi}{\partial z^2} - (\beta^2-1)z\frac{\partial^2\phi}{\partial z^3} - z\frac{\partial^2\Psi}{\partial z^2} - 2\frac{\partial\Psi}{\partial z}, \tag{4.1.17}$$

$$\sigma_{\rho z} = 2\frac{\partial^2\chi}{\partial\rho\partial z} + 2z(\beta^2-1)\frac{\partial^3\phi}{\partial\rho\partial z^2} + 2z\frac{\partial^2\Psi}{\partial\rho\partial z}. \tag{4.1.18}$$

For this solution $\sigma_{\rho z}(\rho,0) = 2(\partial^2\chi/\partial\rho\partial z)_{z=0}$ and $u_z(\rho,0)$ and $\sigma_{zz}(\rho,0)$ are again given by equations (4.1.12) and (4.1.13).

4.2 Solutions Appropriate to a Half-Space

If we are concerned with the half-space $z=0$ we may take

$$\chi \equiv 0, \quad \phi = \mathcal{H}_0[\xi^{-3}\{\beta^{-2}f(\xi) - a(\xi)\}e^{-\xi z}; \xi\to\rho],$$

$$\Psi = -(\beta^2-1)\mathcal{H}_0[\xi^{-2}a(\xi)e^{-\xi z}; \xi\to\rho]$$

to obtain the displacement field

$$u_\rho = -\beta^{-2}\mathcal{H}_0[\xi^{-2}\{f(\xi) - \beta^2 a(\xi) - (\beta^2-1)\xi z f(\xi)\}e^{-\xi z}; \xi\to\rho] \tag{4.2.1}$$

$$u_z = \mathcal{H}_0[\xi^{-2}\{f(\xi) - a(\xi) + (1-\beta^{-2})\xi z f(\xi)\}e^{-\xi z}; \xi\to\rho] \tag{4.2.2}$$

and the temperature field by

$$\Theta = \frac{2(\beta^2-1)}{b}\mathcal{H}_0\left[\xi^{-1}a(\xi)e^{-\xi z};\xi\to\rho\right]. \qquad (4.2.3)$$

The z-components of the stress tensor corresponding to these fields are given by the equations

$$\sigma_{zz} = -2(1-\beta^{-2})\mathcal{H}_0\left[\xi^{-1}f(\xi)(1+\xi z)e^{-\xi z};\xi\to\rho\right], \qquad (4.2.4)$$

$$\sigma_{\rho z} = -2(1-\beta^{-2})z\mathcal{H}_0\left[f(\xi)e^{-\xi z};\xi\to\rho\right] \qquad (4.2.5)$$

so that

$$\sigma_{\rho z}(\rho,0) = 0$$

$$\sigma_{zz}(\rho,0) = -2(1-\beta^{-2})\mathcal{H}_0\left[\xi^{-1}f(\xi);\rho\right] \qquad (4.2.6)$$

$$u_z(\rho,0) = \mathcal{H}\left[\xi^{-2}\{f(\xi)-a(\xi)\};\rho\right] \qquad (4.2.7)$$

$$\Theta(\rho,0) = 2(\beta^2-1)b^{-1}\mathcal{H}_0\left[\xi^{-1}a(\xi);\rho\right]. \qquad (4.2.8)$$

This form of solution is useful in the discussion of the Boussinesq problem for a heated punch; see § 4 below.

On the other hand, the harmonic functions

$$\chi \equiv 0, \quad \phi = \frac{1}{2}\mathcal{H}_0\left[\xi^{-2}\{(\beta^2-1)^{-1}f(\xi)+a(\xi)\}e^{-\xi z};\xi\to\rho\right] \qquad (4.2.9)$$

$$\Psi = \frac{1}{2}\beta^2\mathcal{H}_0\left[\xi^{-1}a(\xi)e^{-\xi z};\xi\to\rho\right] \qquad (4.2.10)$$

lead to a displacement field for which

$$\sigma_{\varrho z}(\varrho,0) = 0$$

$$u_z(\varrho,0) = \frac{1}{2}(1-\beta^{-2})^{-1}\mathcal{H}_0[\xi^{-1}f(\xi);\varrho] \tag{4.2.11}$$

$$\sigma_{zz}(\varrho,0) = \mathcal{H}_0[a(\xi)-f(\xi);] \tag{4.2.12}$$

Also for this solution

$$\theta(\varrho,0) = -\beta^2/b\,\mathcal{H}_0[a(\xi);\varrho], \quad \theta_z(\varrho,0) = \beta^2/b\,\mathcal{H}_0[\xi a(\xi);\varrho]. \tag{4.2.13}$$

This is the form of solution which is most useful in the analysis of crack problems; see § 5 below.

4.3 Thermal Stresses in a Half-Space

We now consider the problem of determining the stress induced in the half-space $z>0$ when the temperature of the boundary $z=0$ is subjected to a prescribed variation (which does not vary with the time), that boundary being free from applied stress. The boundary conditions therefore assume the form

$$\theta(x,y,0) = \theta_0(x,y) \tag{4.3.1}$$

and

$$\sigma_{xz}(x,y,0) = \sigma_{yz}(x,y,0) = \sigma_{zz}(x,y,0) = 0$$

for all $(x,y)\in\mathbb{R}^2$.

If we take (cf. Sneddon & Lockett, 1960)

$$\phi(x,y,z) = -\frac{1}{2}b(\beta^2-1)^{-1}\mathcal{F}^*_{(2)}[\gamma^{-2}\theta_0^*(\xi,\zeta)e^{-\gamma z};(\xi,\zeta)\to(x,y)],$$

in which $\gamma = |(\xi^2+\zeta^2)^{1/2}|$ and

$$\theta_0^*(\xi,\zeta) = \mathcal{F}_{(2)}[\theta_0(x,y);\,(x,y)\to(\xi,\zeta)]\,, \qquad (4.3.2)$$

in the Lur'e solution (4.1.14) we obtain the solution

$$u_x = \frac{1}{2}b(\beta^2-1)^{-1} i\mathcal{F}_{(2)}^*[\xi\gamma^{-2}\theta_0^*(\xi,\zeta)e^{-\gamma z};\,(\xi,\zeta)\to(x,y)]\,, \qquad (4.3.3)$$

$$u_y = \frac{1}{2}b(\beta^2-1)^{-1} i\mathcal{F}_{(2)}^*[\zeta\gamma^{-2}\theta_0^*(\xi,\zeta)e^{-\gamma z};\,(\xi,\zeta)\to(x,y)]\,, \qquad (4.3.4)$$

$$u_z = \frac{1}{2}b(\beta^2-1)^{-1} \mathcal{F}_{(2)}^*[\gamma^{-1}\theta_{(2)}^*(\xi,\zeta)e^{-\gamma z};\,(\xi,\zeta)\to(x,y)]\,, \qquad (4.3.5)$$

$$\theta = \mathcal{F}_{(2)}^*[\theta_0^*(\xi,\zeta)e^{-\gamma z};\,(\xi,\zeta)\to(x,y)]\,. \qquad (4.3.6)$$

In this case we have

$$\sigma_{xz} = \sigma_{yz} = \sigma_{zz} = 0 \qquad (4.3.7)$$

at all points of the body. This is the solution first derived by Sternberg and McDovell (1957); is shows that the stress field due to arbitrary surface temperatures is plane and parallel to the boundary.

In the case in which the prescribed surface temperature $\theta_0(x,y)$ is a function of $\varrho = (x^2+y^2)^{1/2}$ only, we take

$$a(\xi) = f(\xi) = -b^{-1}\beta^2\bar{\theta}_0(\xi)$$

with

$$\bar{\theta}_0(\xi) = \mathcal{H}_0[\theta_0(\varrho);\,\xi] \qquad (4.3.8)$$

in equations (4.2.9), (4.2.10) to obtain the solution

$$u_\varrho(\varrho, z) = \frac{1}{2} b(\beta^2-1)^{-1} \mathcal{H}_1[\xi^{-1}\bar{\Theta}_0(\xi)e^{-\xi z}; \xi \to \varrho], \qquad (4.3.9)$$

$$u_z(\varrho, z) = \frac{1}{2} b(\beta^2-1)^{-1} \mathcal{H}_0[\xi^{-1}\bar{\Theta}_0(\xi)e^{-\xi z}; \xi \to \varrho], \qquad (4.3.10)$$

$$\Theta(\varrho, z) = \mathcal{H}_0[\bar{\Theta}_0(\xi)e^{-\xi z}; \xi \to \varrho]. \qquad (4.3.11)$$

As a special case of the use of these formulae, we consider the problem which is considered in some detail by Sternberg and McDowell, namely that in which

$$\Theta_0(\varrho) = \Theta_0 H(1-\varrho) \qquad (4.3.12)$$

with Θ_0 a constant. For this temperature distribution

$$\bar{\Theta}_0(\xi) = \Theta_0 \xi^{-1} J_1(\xi)$$

and in the notation of Eason et. al. (1955) we have

$$u_\varrho(\varrho, z) = \frac{1}{2} b(\beta^2-1)^{-1} \Theta_0 J(1,1;-1) \qquad (4.3.13)$$

$$u_z(\varrho, z) = -\frac{1}{2} b(\beta^2-1)^{-1} \Theta_0 J(1,0;-1) \qquad (4.3.14)$$

where

$$J(\mu,\nu;\lambda) = \int_0^\infty J_\mu(\xi) J_\nu(\varrho\xi) e^{-\xi z} \xi^\lambda d\xi. \qquad (4.3.15)$$

The integrals $J(1,1;-1)$ and $J(1,0;-1)$ have been tabulated for ranges of values of ϱ and z in Tables 9, 10 and 11 on pp. 545-546 of Eason et al. (1955), so that it is an easy matter to calculate the components of the displacement vector at any point. For example, Fig. 1 shows the variation of the displacement component, u_z, of the displacement in planes parallel to the boundary.

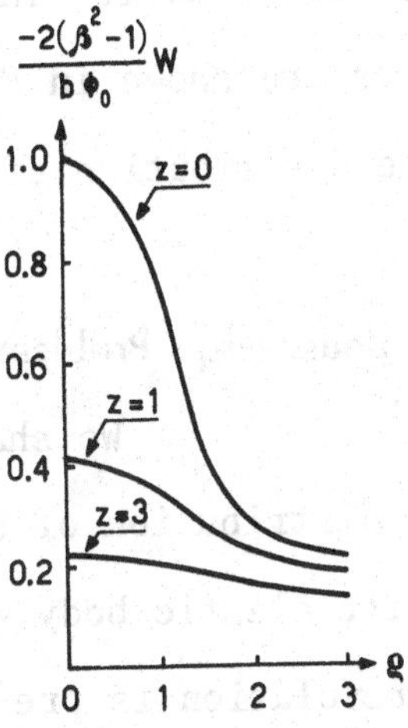

Fig.1

It is of interest also to calculate the difference of principal stresses

$$\tau = \varrho \frac{\partial}{\partial \varrho}(\varrho^{-1} u_\varrho) . \qquad (4.3.16)$$

It is easily shown that in the general case

$$\tau = b(\beta^2-1)^{-1} \int_0^\infty \xi \bar{\Theta}_0(\xi)[J_0(\varrho\xi) - 2J_1(\varrho\xi)/(\varrho\xi)] e^{-\xi z} d\xi \qquad (4.3.17)$$

and in the case in which $\Theta(x,y,0) = \Theta_0 H(1-\varrho)$,

$$\tau = (\beta^2-1)^{-1}\Theta_0[J(1,0;0) - (2/\varrho)J(1,1;-1)] . \qquad (4.3.18)$$

The values of τ at a grid of points in the ϱz-plane can be calculated easily from Tables 6 and 11 of Eason et al. and is then a simple matter to draw the lines joining points with the same value of this stress difference. The resulting set of curves will be

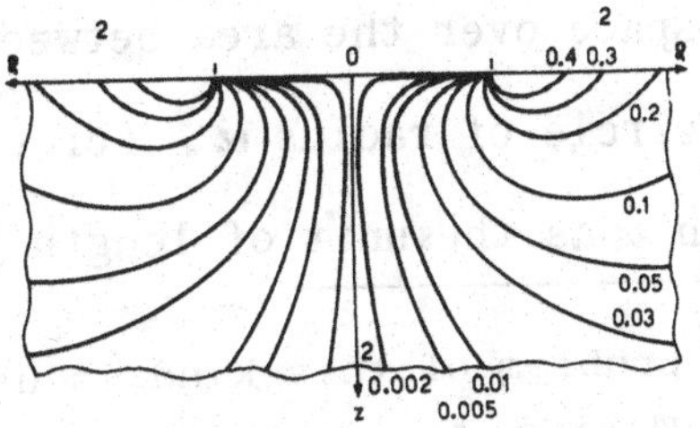

Fig. 2. Section by the plane θ = *constant* of the surface $|\sigma_\varrho - \sigma_\theta|$ = *constant*, in a semi-infinite solid. The numbers refer to values of $(\beta^2-1)\cdot(\sigma_\theta - \sigma_\varrho)/b\phi_0$.

the curves obtained by cutting surfaces of equal maximum shearing stress by a plane through the z -axis, and will correspond to the isochromatic lines of two-dimensional elasticity. These contours are shown in Fig. 2 for the simple case in which the surface distribution of temperature is given by equation (4.3.12).

4.4 Boussinesq's Problem for a Heated Punch

We shall now consider the problem of determining the distribution of stress in a homogeneous, isotropic, semi-infinite elastic body when a rigid punch in the form of a surface of revolution is pressed against it. (*) We assume that the rigid punch is heated and so produces a non-uniform distribution of temperature on the free surface of the elastic solid.

We suppose that the axis of the punch is normal to the boundary plane of the elastic solid. If we take the undisturbed boundary to be the plane $z = 0$ and the point at which the tip of the punch begins to ident the solid to be the origin of coordinates. Since the punch is a solid of revolution and it is pressed normally against the boundary, it will fit the elastic half-space over the area between the apex of the punch and a certain circle of radius α. For convenience, we shall take this length α as the unit of length in our calculations.

(*) A problem of this kind is usually called a *Boussinesq problem* since the isothermal problem was first discussed in detail by J. Boussinesq. (See, e.g., pp 202-255, 713-719 of Boussinesq, 1885).

We assume that the bodies in contact are smooth so that

$$\sigma_{\varrho z}(\varrho,0) = 0 , \quad \varrho \geq 0 \tag{4.4.1}$$

and that the z-component of the displacement is specified over the disk $0 \leq \varrho \leq 1$, $z = 0$, while outside that region we impose the condition of zero normal stress. Thus the remaining boundary conditions on the plane $z = 0$ are the mixed conditions

$$u_z(\varrho,0) = g(\varrho), \quad 0 \leq p \leq 1 \tag{4.4.2}$$

$$\sigma_{zz}(\varrho,0) = 0 , \quad \varrho > 1 . \tag{4.4.3}$$

We shall also assume that the heating of the punch is also symmetric and that the temperature field is of the form $T_0\{1+\Theta(\varrho,z)\}$.

If we take a displacement field given by the equations (4.2.1) and (4.2.2) we see from equations (4.2.6) through (4.2.8) that $a(\xi)$ is determined by the equation

$$a(\xi) = \frac{1}{2}b(\beta^2-1)^{-1}\xi\mathcal{H}_0[\Theta(\varrho,0);\xi] \tag{4.4.4}$$

and $f(\xi)$ by the pair of dual integral equations

$$\mathcal{H}_0[\xi^{-2}f(\xi);\varrho] = A(\varrho)+g(\varrho), \quad 0 \leq \varrho \leq 1 \tag{4.4.5}$$

$$\mathcal{H}_0[\xi^{-1}f(\xi);\varrho] = 0, \quad \varrho > 1 \tag{4.4.6}$$

where the function $A(\varrho)$ is defined by the equation

$$A(\varrho) = \mathcal{H}_0[\xi^{-2}a(\xi);\varrho] . \tag{4.4.7}$$

We write the solution of this pair of dual integral equations in the form

$$f(\xi) = f_1(\xi) + f_2(\xi) \tag{4.4.8}$$

where $f_1(\xi)$ is the solution of the pair

$$\mathcal{H}_0[\xi^{-2} f_1(\xi); \varrho] = A(\varrho), \quad 0 \leq \varrho \leq 1 \tag{4.4.9}$$

$$\mathcal{H}_0[\xi^{-1} f_1(\xi); \varrho] = 0, \quad \varrho > 1 \tag{4.4.10}$$

and $f_2(\xi)$ that of the pair

$$\mathcal{H}_0[\xi^{-2} f_2(\xi); \varrho] = g(\varrho), \quad 0 \leq \varrho \leq 1 \tag{4.4.11}$$

$$\mathcal{H}_0[\xi^{-1} f_2(\xi); \varrho] = 0, \quad \varrho > 1 . \tag{4.4.12}$$

If we take

$$\begin{aligned} f_1(\xi) &= \frac{2\xi}{\pi} \int_0^1 \cos(\xi t) \Psi_1'(t) dt \\ &= \frac{2}{\pi} \sin \xi \, \Psi_1'(1) - \frac{2}{\pi} \int_0^1 \sin(\xi t) \Psi_1''(t) dt \end{aligned} \tag{4.4.13}$$

we find from equation (A.42) that equation (4.4.10) is automatically satisfied and from equation (A.40) that equation (4.4.9) is satisfied if $\Psi_1'(t)$ is such that

$$\mathcal{A}_1[\Psi_1'(t); \varrho] = \sqrt{\left(\frac{1}{2}\pi\right)} A(\varrho), \quad 0 \leq \varrho \leq 1 .$$

Using the inversion theorem (A.30) we deduce that

$$\Psi_1(t) = \int_0^t \frac{\varrho A(\varrho) d\varrho}{\sqrt{(t^2 - \varrho^2)}} . \tag{4.4.14}$$

Now if we write

$$\bar{\Theta}_0(\xi) = \mathcal{H}_0[\Theta(\rho,0)\,;\xi]$$

we see from equations (4.4.4), (4.4.7) and (4.4.14) that

$$\psi_1(t) = \frac{1}{2}b(\beta^2-1)^{-1}\int_0^\infty \xi\bar{\Theta}_0(\xi)d\xi\int_0^t \frac{\rho J_0(\xi\rho)d\rho}{\sqrt{(t^2-\rho^2)}} .$$

Since

$$\int_0^t \frac{\rho J_0(\xi\rho)d\rho}{\sqrt{(t^2-\rho^2)}} = \frac{\sin(\xi t)}{\xi}$$

we find that

$$\psi_1'(t) = \frac{1}{2}b(\beta^2-1)^{-1}\sqrt{\left(\frac{1}{2}\pi\right)}\Theta_c(t) \qquad (4.4.15)$$

where

$$\Theta_c(t) = \mathcal{F}_c\left[\bar{\Theta}_0(\xi)\,;t\right] .$$

From equation (4.4.3) we find that

$$\Theta_c(t) = \mathcal{A}_2\left[\rho\Theta(\rho,0)\,;t\right] . \qquad (4.4.16)$$

Hence we find that

$$f_1(\xi) = b(\beta^2-1)^{-1}(2\pi)^{-1/2}\xi\int_0^1 \Theta_c(t)\cos(\xi t)dt \qquad (4.4.17)$$

where $\Theta_c(t)$ is given by equation (4.4.16).

Similarly, we can show that

$$f_2(\xi) = \frac{2\xi}{\pi}\int_0^1 \cos(\xi t)\Psi_2'(t)dt \tag{4.4.18}$$

with

$$\Psi_2(t) = \int_0^t \frac{\varrho g(\varrho)d\varrho}{\sqrt{(t^2-\varrho^2)}} . \tag{4.4.19}$$

In any specific problem it is of interest to calculate the total load which must be applied to the punch to maintain the prescribed displacement. The load is given by the formula

$$P = -2\pi\int_0^1 \varrho\sigma_{zz}(\varrho,0)d\varrho .$$

From equation (4.2.6) we deduce that

$$P = 4\pi(1-\beta^{-2})\int_0^\infty \xi^{-1}f(\xi)J_1(\xi)d\xi .$$

In the isothermal case the corresponding load is P_0 , where

$$P_0 = 4\pi(1-\beta^{-2})\int_0^\infty \xi^{-1}f_2(\xi)J_1(\xi)d\xi$$

so that

$$\frac{P-P_0}{P_0} = \frac{I_1}{I_2} \tag{4.4.20}$$

where

$$I_j = \int_0^\infty \xi^{-1}f_j(\xi)J_1(\xi)d\xi , \quad (j=1,2) .$$

Substituting the expressions (4.4.13), (4.4.15) for $f_1(\xi)$ and (4.4.18) for $f_2(\xi)$ and making use of the integral

$$\int_0^\infty J_1(\xi)\cos(\xi t)d\xi = 1, \quad 0 < t < 1$$

we find that

$$\frac{P-P_0}{P_0} = \frac{1}{2}b(\beta^2-1)^{-1}\sqrt{\left(\frac{1}{2}\pi\right)}\Psi_2^{-1}(1)\int_0^1 \theta_c(t)dt . \qquad (4.4.21)$$

Special Types of Temperature Fields.

We now consider, briefly, the functions corresponding to certain types of temperature field.

Type (a): First of all take the case in which the temperature field is created by the surface conditions

$$\theta(\rho,0) = \theta_1(\rho)H(1-\rho) .$$

For this surface temperature we have

$$\bar{\theta}_0(\xi) = \int_0^1 \rho\theta_1(\rho)J_0(\xi\rho)d\rho \qquad (4.4.22)$$

so that, interchanging the order of the integrations in the definition of the function $\theta_c(t)$, we obtain expression

$$\theta_c(t) = \frac{2}{\pi}\int_t^1 \frac{s\theta_1(s)ds}{\sqrt{(s^2-t^2)}} \qquad (4.4.23)$$

from which we deduce by means of equation (4.4.17) that

$$f_1(\xi) = \frac{1}{2}b(\beta^2-1)^{-1}\xi\bar{\theta}_0(\xi) . \qquad (4.4.24)$$

We also deduce that

$$\int_0^1 \Theta_c(t)dt = \sqrt{\left(\frac{1}{2}\pi\right)} \int_0^1 \varrho\Theta_1(\varrho)d\varrho$$

so that

$$\frac{P-P_0}{P_0} = \frac{1}{4}b\pi(\beta^2-1)^{-1}(J_1/J_2) \tag{4.4.25}$$

where

$$J_1 = \int_0^1 \varrho\Theta_1(\varrho)d\varrho\ , \quad J_2 = \int_0^1 \frac{\varrho g(\varrho)d\varrho}{\sqrt{(1-\varrho^2)}}\ . \tag{4.4.26}$$

Type (b): Next we discuss the temperature field produced by the boundary conditions

$$\frac{\partial\Theta}{\partial z} = -Q(\varrho)\ , \quad 0 \leq \varrho \leq 1$$

$$\Theta = 0\ , \quad \varrho > 1\ .$$

For this surface temperature distribution we find that $\bar{\Theta}_0(\xi)$ satisfies the dual integral equations

$$\mathcal{H}_0[\xi\bar{\Theta}_0(\xi);\varrho] = Q(\varrho)\ , \quad 0 \leq \varrho \leq 1$$

$$\mathcal{H}_0[\bar{\Theta}_0(\xi);\varrho] = 0\ , \quad \varrho > 1$$

which have the solution

$$\bar{\Theta}_0(\xi) = \frac{2}{\pi}\xi^{-1}\int_0^1 q(t)\sin(\xi t)dt \tag{4.4.27}$$

where

$$q(t) = \int_0^t \frac{\rho Q(\rho)d\rho}{\sqrt{(t^2-\rho^2)}} . \tag{4.4.28}$$

Hence we deduce that

$$\Theta_c(t) = \frac{2}{\pi}\int_t^1 q(u)du \tag{4.4.29}$$

and that

$$\int_0^1 \Theta_c(t)dt = \frac{2}{\pi}\int_0^1 uq(u)du . \tag{4.4.30}$$

We also have from (4.4.17) that

$$f_1(\xi) = b\pi^{-1}(\beta^2-1)^{-1}\int_0^1 q(u)\sin(\xi u)du . \tag{4.4.31}$$

Using the result (4.4.30) in (4.4.21) we deduce that in this case $(P-P_0)/P_0$ is again given by equation (4.4.25) with J_2 defined by the second of equations (4.4.26) but J_1 now defined by

$$J_1 = \frac{2}{\pi}\int_0^1 uq(u)du . \tag{4.4.32}$$

The special case $Q(\rho)=Q_0$, a constant, has

$$\bar{\Theta}_0(\xi) = (2/\pi)Q_0\xi^{-3}(\sin\xi - \xi\cos\xi) ,$$

$$\Theta_c(t) = (2\pi)^{-1/2}Q_0(1-t^2), \quad 0<t<1 ,$$

$$f_1(\xi) = b(\beta^2-1)^{-1}Q_0\xi^{-2}(\sin\xi - \xi\cos\xi) ,$$

$$\frac{P-P_0}{P_0} = bQ_0/\{6(\beta^2-1)J_2\} .$$

Case (c). If the temperature field is due to the boundary conditions

$$\theta = \theta_2(\varrho), \quad 0 \leq \varrho \leq 1,$$

$$\frac{\partial \theta}{\partial z} = 0, \quad \varrho > 1,$$

the function $\bar{\theta}_0(\xi)$ is the solution of the dual integral equations

$$\mathcal{H}_0[\bar{\theta}_0(\xi); \varrho] = \theta_2(\varrho), \quad 0 \leq \varrho \leq 1,$$

$$\mathcal{H}_0[\xi\bar{\theta}_0(\xi); \varrho] = 0, \quad \varrho > 1.$$

These are of the same type as (4.4.9) and (4.4.10) and therefore have the solution

$$\bar{\theta}_0(\xi) = \frac{2}{\pi}\xi^{-1}\int_0^1 \Psi_3'(t)\cos(\xi t)dt,$$

where

$$\Psi_3(t) = \int_0^t \frac{s\theta_2(s)ds}{\sqrt{(t^2 - s^2)}}.$$

It follows from equation (4.4.4) that

$$a(\xi) = b(\beta^2 - 1)^{-1}\pi^{-1}\int_0^1 \Psi_3'(t)\cos(\xi t)dt.$$

The use of these formulae in the calculation of the displacement and stress fields in the cases in which the punch is a flat-ended circular cylinder or a cone is illustrated by George and Sneddon (1962) to which the reader is referred for details.

4.5 The Thermoelastic Problem for a Penny-Shaped Crack

We now consider the problem of determining the distribution of stress in the vicinity of the penny-shaped crack $0 \leq \rho \leq 1$, $z = 0$, when the temperature on the crack surface is a prescribed function $q(\rho)$ of ρ. (Cf. Olesiak & Sneddon, 1960; Tweed, 1969). If the crack is free from stress, we have the boundary conditions

$$\theta(\rho, 0) = q(\rho), \quad (0 \leq \rho \leq 1), \tag{4.5.1}$$

$$\theta_z(\rho, 0) = 0, \quad (\rho > 1), \tag{4.5.2}$$

$$\sigma_{zz}(\rho, 0) = 0, \quad (0 \leq \rho \leq 1), \tag{4.5.3}$$

$$u_z(\rho, 0) = 0, \quad (\rho > 1), \tag{4.5.4}$$

by means of which we may determine the temperature and displacement fields in the half-space $z \geq 0$ and hence, by symmetry, in the whole space.

The solution is given by equations (4.1.6), (4.1.7) with χ, ϕ, Ψ given by equations (4.2.9) and (4.2.10) provided that $\alpha(\xi)$, $f(\xi)$ are chosen to satisfy

$$\mathcal{H}_0[a(\xi); \rho] = -\beta^{-2} bq(\rho), \quad (0 \leq \rho \leq 1) \tag{4.5.5}$$

$$\mathcal{H}_0[\xi a(\xi); \rho] = 0, \quad (\rho > 1) \tag{4.5.6}$$

$$\mathcal{H}_0[a(\xi) - f(\xi); \rho] = 0, \quad (0 \leq \rho \leq 1) \tag{4.5.7}$$

$$\mathcal{H}_0[\xi^{-1} f(\xi); \rho] = 0, \quad (\rho > 1). \tag{4.5.8}$$

If we make the representation

$$\left(\frac{1}{2}\pi\right)^{1/2} a(\xi) = -\xi^{-1}\int_0^1 g(t)\cos(\xi t)dt = \xi^{-2}\int_0^1 g'(t)\sin(\xi t)dt \tag{4.5.9}$$

it follows from equation (A.42) that equation (4.5.6) is automatically satisfied and from equation (A.39) that

$$\mathcal{A}_1[g(t); \rho] = \beta^{-2} bq(\rho), \quad 0 < \rho < 1. \tag{4.5.10}$$

Using equation (A.30) we find that

$$g(t) \equiv \beta^{-2} bD_t \mathcal{A}_1[\rho q(\rho); t]. \tag{4.5.11}$$

Similarly equation (4.5.8) is automatically satisfied by the representation

$$\left(\frac{1}{2}\pi\right)^{1/2} f(\xi) = -\int_0^1 h(t)\sin(\xi t)dt, \quad h(0) = 0. \tag{4.5.12}$$

Using (4.5.5) we can rewrite (4.5.7) as

$$\mathcal{H}_0[f(\xi); \rho] = -\beta^{-2} bq(\rho), \quad 0 \leq \rho \leq 1$$

so that $h(t)$ must be chosen to be such that for $0 \leq t \leq 1$

$$\mathcal{A}_1[h'(t); \rho] = \beta^{-2} bq(\rho), \quad 0 \leq \rho \leq 1.$$

Making use of (A.30) and the condition $h(0) = 0$ we deduce that

$$h(t) \equiv \beta^{-2} b\mathcal{A}_1[\rho q(\rho); t], \quad 0 \leq \rho \leq 1. \tag{4.5.13}$$

From equations (4.2.11), (4.5.12) we deduce that

$$u_z(\rho,0) = \frac{\beta^2}{\sqrt{(2\pi)}(\beta^2-1)} \int_\rho^1 \frac{h(t)\,dt}{\sqrt{(t^2-\rho^2)}}, \quad 0 \leq \rho \leq 1. \quad (4.5.14)$$

For example, if the faces of the crack are kept at a temperature Θ_0 below the reference temperature so that

$$q(\rho) = -\Theta_0$$

it follows that

$$\mathcal{A}_1[\rho q(\rho);t] = -(2/\pi)^{1/2}\Theta_0 t$$

and hence that

$$g(t) = -(2/\pi)^{1/2} b\beta^{-2}\Theta_0, \quad h(t) = -(2/\pi)^{1/2} b\beta^{-2}\Theta_0 t.$$

From equation (4.5.14) we deduce that

$$u_z(\rho,0) = \frac{b\Theta_0}{\pi(\beta^2-1)}(1-\rho^2)^{1/2}. \quad (4.5.15)$$

Also

$$a(\xi) = \frac{2b\Theta_0}{\pi\beta^2}\frac{\sin\xi}{\xi^2}, \quad f(\xi) = \frac{2b}{\pi\beta^2}\Theta_0\left[\frac{\sin\xi}{\xi^2} - \frac{\cos\xi}{\xi}\right]$$

and so

$$\sigma_{zz}(\rho,0) = (2b\Theta_0/\pi\beta^2)\mathcal{H}_0[\xi^{-1}\cos\xi;\rho]$$

showing that

$$\sigma_{zz}(\rho,0) = (2b\Theta_0/\pi\beta^2)(\rho^2-1)^{-1/2}, \qquad (\rho>1). \quad (4.5.16)$$

We have similar results when the flux of heat across the crack surfaces is prescribed. The mixed boundary conditions (4.5.1), (4.5.2) are replaced by the single condition

$$\Theta_z(\rho,0) = Q(\rho)H(1-\rho). \quad (4.5.17)$$

Substituting this condition in the second of the equations (4.2.13) and using the Hankel inversion theorem we find that

$$a(\xi) = b\beta^{-2}\xi^{-1}\int_0^1 rQ(r)J_0(\xi r)dr \quad (4.5.18)$$

from which it follows by the first equation of the pair (4.2.13) that

$$q(\rho) \equiv \Theta(\rho,0) = -\mathcal{H}_0\left[\xi^{-1}\int_0^1 rQ(r)J_0(\xi r)dr;\rho\right]$$

and

$$\mathcal{A}_1[\rho q(\rho);t] = \int_0^1 \mathcal{A}_1[\rho J_0(\xi\rho);\rho\rightarrow t]d\xi\int_0^1 rQ(r)J_0(\xi r)dr.$$

Now

$$\mathcal{A}_1[\rho J_0(\xi\rho);\rho\rightarrow t] = (2/\pi)^{1/2}\xi^{-1}\sin(\xi t)$$

so that

$$\mathcal{A}_1[\rho q(\rho);t] = (2/\pi)^{1/2}\int_0^1 rQ(r)dr\int_0^\infty \xi^{-1}J_0(\xi r)\sin(\xi t)d\xi.$$

From equation (4.5.13) we deduce that

$$h(t) = (2/\pi)^{1/2}\beta^{-2}b\int_0^1 rQ(r)dr\int_0^\infty \xi^{-1}J_0(\xi r)\sin(\xi t)d\xi$$

and from (4.5.11) that

$$g(t) = (2/\pi)^{1/2}\beta^{-2}b\int_0^1 rQ(r)dr\int_0^\infty J_0(\xi r)\cos(\xi t)d\xi .$$

Making use of the integrals

$$\int_0^\infty J_0(\xi r)\cos(\xi t)d\xi = (r^2-t^2)^{-1/2}H(r-t) ,$$

$$\int_0^\infty \xi^{-1}J_0(\xi r)\sin(\xi t)d\xi = \sin^{-1}(t/r)H(r-t) + \frac{1}{2}\pi H(t-r)$$

$(r>0, t>0)$ we obtain the formulae

$$h(t) = (2/\pi)^{1/2}\beta^{-2}b\int_t^1 rQ(r)\sin^{-1}\left(\frac{t}{r}\right)dr + \left(\frac{1}{2}\pi\right)^{1/2}\beta^{-2}b\int_0^t rQ(r)dr$$

$$g(t) = (2/\pi)^{1/2}\beta^{-2}b\int_t^1 \frac{rQ(r)dr}{\sqrt{(r^2-t^2)}} .$$

For example, if

$$Q(\rho) = Q_0 ,$$

a constant,

$$h(t) = (2\pi)^{-1/2}\beta^{-2}bQ_0\{\sin^{-1}t + t\sqrt{(1-t^2)}\}$$

$$g(t) = (2/\pi)^{1/2}\beta^{-2}bQ\sqrt{(1-t^2)} .$$

The calculation of the stress and displacement fields is illustrated in Olesiak and Sneddon (1960). Attention should also be drawn to the papers by B.R. Das (1958, 1959) where the above method is generalized to provide the solution of thermoelastic problems concerning a long circular cylinder containing a penny-shaped crack.

B.R. Das (1971) has also considered the distribution of thermal stress in the neighbourhood of an *external* crack in an infinite solid. In this case the equations (4.5.1) - (4.5.4) are replaced by the boundary conditions

$$\theta_z(\rho,0) = 0\,, \quad 0<\rho<1\,,$$

$$\theta(\rho,0) = q(\rho)\,, \quad \rho>1\,,$$

$$u_z(\rho,0) = 0\,, \quad 0<\rho<1\,,$$

$$\sigma_{zz}(\rho,0) = 0\,, \quad \rho>1$$

and representations of the type (4.5.9) are replaced by expressions of the form

$$a(\xi) = (2/\pi)^{1/2}\xi^{-1}\int_1^\infty g(t)\sin(\xi t)dt\,.$$

4.6 Thermal Stresses in a Thick Plate

We now consider some simple problems concerning the stress field in a thick plate. For convenience we take the plate to be defined by the inequalities $-1 \leq z \leq 1$; in other words, all lengths are measured as ratios of half the thickness of the plate. We suppose that the surfaces of the plate are kept at prescribed temperatures, i.e. we assume that

$$\Theta(x,y,+1) = f(x,y)\,, \quad \Theta(x,y,-1) = g(x,y) \tag{4.6.1}$$

where the functions f and g are prescribed.

If we take

$$\phi(x,y,z) = -\frac{1}{2}b(\beta^2-1)^{-1}\mathfrak{F}^*_{(2)}[\zeta^{-2}\operatorname{cosech}2\zeta\{f^*\sinh(1+z)\zeta + g^*\sinh(1-z)\zeta\};(\xi,\eta)\rightarrow(x,y)] \tag{4.6.2}$$

in which $\zeta^2 = \xi^2+\eta^2$ and

$$f^* = \mathfrak{F}_{(2)}[f(x,y);(x,y)\rightarrow(\xi,\eta)], \quad g^* = \mathfrak{F}_{(2)}[g(x,y);(x,y)\rightarrow(\xi,\eta)]$$

in the Lur'e solution (4.1.14) we find that

$$u_x = +\frac{1}{2}b(\beta^2-1)\mathfrak{F}^*_{(2)}[i\xi\zeta^{-2}\operatorname{cosech}(2\zeta)\{f^*\sinh(1+z)\zeta + g^*\sinh(1-z)\zeta\}] \tag{4.6.3a}$$

$$u_y = +\frac{1}{2}b(\beta^2-1)\mathfrak{F}^*_{(2)}[i\eta\zeta^{-2}\operatorname{cosech}(2\zeta)\{f^*\sinh(1+z)\zeta + g^*\sinh(1-z)\zeta\}] \tag{4.6.3b}$$

$$u_z = +\frac{1}{2}b(\beta^2-1)\mathfrak{F}^*_{(2)}[\zeta^{-1}\operatorname{cosech}(2\zeta)\{f^*\cosh(1+z)\zeta - g^*\cosh(1-z)\zeta\}] \tag{4.6.3c}$$

$$\Theta = \mathfrak{F}^*_{(2)}[\operatorname{cosech}2\zeta\{f^*\sinh(1+z)\zeta + g^*\sinh(1-z)\zeta\}] \tag{4.6.3d}$$

and it is easily verified from this last equation that the boundary conditions (4.6.1) are satisfied.

Similarly, if, in the Lur'e solution, we take

$$\phi(x,y,z) = -\frac{1}{2}b(\beta^2-1)^{-1}\mathfrak{F}^*_{(2)}[\zeta^{-2}\operatorname{sech}\zeta f^*\cosh\zeta z;(\xi,\eta)\rightarrow(x,y)] \tag{4.6.4}$$

we obtain the solution

$$u_x = \frac{1}{2}b(\beta^2-1)^{-1}\mathfrak{F}^*_{(2)}[i\xi\zeta^{-2}\operatorname{sech}\zeta f^*\cosh\zeta z] \tag{4.6.5a}$$

$$u_y = \frac{1}{2}b(\beta^2-1)^{-1}\mathfrak{F}^*_{(2)}[i\eta\zeta^{-1}\operatorname{sech}\zeta f^*\cosh(\zeta z)] \tag{4.6.5b}$$

$$u_z = \frac{1}{2}b(\beta^2-1)^{-1}\mathfrak{F}^*_{(2)}[\zeta^{-1}\operatorname{sech}\zeta f^*\sinh(\zeta z)] \tag{4.6.5c}$$

$$\theta = \mathfrak{F}^*_{(2)}[\operatorname{sech}\zeta f^*\cosh(\zeta z)] . \tag{4.6.5d}$$

This is obviously the solution for the plate $-1 \leqslant z \leqslant 1$ when

$$\theta(x,y,\pm 1) = f(x,y) \tag{4.6.6}$$

but it is also the solution for the plate $0 \leqslant z \leqslant 1$ when

$$\theta(x,y,1) = f(x,y) \tag{4.6.7}$$

$$\frac{\partial\theta(x,y,0)}{\partial z} = 0 . \tag{4.6.8}$$

It should also be observed that, for this solution

$$u_z(x,y,0) = 0 . \tag{4.6.9}$$

For these solutions $\sigma_{xz} = \sigma_{yz} = \sigma_{zz} = 0$ at every point of the plate so that the Sternberg-McDowell result that the stress field induced by an arbitrary distribution of surface temperature is plane and parallel to the boundary holds for a thick plate as well as for a half-space.

For problems with axial symmetry we use a Hankel transform solution. For example, if the boundary conditions (4.6.1) take the form

$$\theta(\rho,+1) = f(\rho), \quad \theta(\rho,-1) = g(\rho), \tag{4.6.10}$$

then instead of (4.6.2) we use the representation

$$\phi(\rho,z) = -\frac{1}{2}b(\beta^2-1)^{-1}\mathscr{H}_0[\xi^{-2}\operatorname{cosech}2\xi\{\bar{f}\sinh(1+z)\xi + \bar{g}\sinh(1-z)\xi\};\xi\to\rho] \tag{4.6.11}$$

where $\bar{f}(\xi) = \mathscr{H}_0[f(\rho);\xi]$, $\bar{g}(\xi) = \mathscr{H}_0[g(\rho);\xi]$. In this case the displacement and temperature fields are given by the equations

$$u_\rho = \frac{1}{2}b(\beta^2-1)^{-1}\mathscr{H}_1[\xi^{-1}\operatorname{cosech}2\xi\{\bar{f}\sinh(1+z)\xi+\bar{g}\sinh(1-z)\xi\};\xi\to\rho] \tag{4.6.12a}$$

$$u_z = \frac{1}{2}b(\beta^2-1)^{-1}\mathscr{H}_0[\xi^{-1}\operatorname{cosech}2\xi\{\bar{f}\cosh(1+z)\xi-\bar{g}\cosh(1-z)\xi\};\xi\to\rho] \tag{4.6.12b}$$

$$\theta = \mathscr{H}_0[\operatorname{cosech}2\xi\{\bar{f}\sinh(1+z)\xi + \bar{g}\sinh(1-z)\xi\};\xi\to\rho]. \tag{4.6.12c}$$

Similarly to obtain the solution in $0 \leq z \leq 1$ corresponding to the boundary conditions

$$\theta(p,1) = f(\rho) \tag{4.6.13}$$

$$\frac{\partial \Theta(\varrho,0)}{\partial z} = 0 \tag{4.6.14}$$

we take

$$\phi(\varrho,z) = -\frac{1}{2} b(\beta^2-1)^{-1} \mathcal{H}_0[\xi^{-2} \operatorname{sech}\xi \bar{f}(\xi) \cosh\xi z\,;\,\xi\to\varrho]. \tag{4.6.15}$$

This gives

$$u_\varrho = \frac{1}{2} b(\beta^2-1)^{-1} \mathcal{H}_1[\xi^{-1} \operatorname{sech}\xi \bar{f}(\xi) \cosh\xi z\,;\,\xi\to\varrho] \tag{4.6.16a}$$

$$u_z = \frac{1}{2} b(\beta^2-1)^{-1} \mathcal{H}_0[\xi^{-1} \operatorname{sech}\xi \bar{f}(\xi) \sinh\xi z\,;\,\xi\to\varrho] \tag{4.6.16b}$$

$$\Theta = \mathcal{H}_0[\bar{f}(\xi) \operatorname{sech}\xi \cosh\xi z\,;\,\xi\to\varrho]. \tag{4.6.16c}$$

For any given distributions of temperature on the surfaces of the plate. We can calculate the zero-order Hankel transforms $\bar{f}(\xi)$, $\bar{g}(\xi)$ and inserting them in equations (4.6.12) or (4.6.16) calculate the temperature and displacement fields within the plate. In the general case the evaluation of these integrals would be difficult, because of the occurrence of the factor cosech 2ξ in the integrand. By suitably choosing the functions $f(\varrho)$ and $g(\varrho)$ we can, however, obtain integrals which can easily be evaluated, and obtain the solution of a representative problem.

For example if in equations (4.6.10) we take

$$f(\varrho) = [(k^2-4)^2\Theta_0/2k]\{(k-2)[\varrho^2+(k-2)^2]^{-3/2} - (k+2)[\varrho^2+(k+2)^2]^{-3/2}\}$$

$(k > 2)$, and (4.6.17)

$$g(\varrho) = 0 ,$$

then

$$\bar{f}(\xi) = \left[\frac{1}{4}(k^2-4)^2\Theta_0/k\right]e^{-k\xi}\sinh(2\xi), \quad (k>2) \qquad (4.6.18)$$

$$\bar{g}(\xi) = 0 .$$

The temperature field is then given by equation (4.6.12c) in the form

$$\Theta(\varrho,z) = [(k^2-4)^2\Theta_0/8k]\{(k-1-z)[\varrho^2+(k-1-z)^2]^{-3/2} - (k+1+z)[\varrho^2+(k+1+z)^2]^{-3/2}\}.$$

Similarly from equations (4.6.12a) (4.6.12b) we deduce that

$$u_\varrho = \varrho^{-1}\theta\{(k+1+z)[\varrho^2+(k+1+z)^2]^{-1/2} - (k-1-z)[\varrho^2+(k-1-z)^2]^{-1/2}\}$$

$$u_z = \theta\{[\varrho^2+(k-1-z)^2]^{-1/2} + [\varrho^2+(k+1+z)^2]^{-1/2}\}$$

where θ is the constant defined by the equation

$$\theta = \frac{(k^2-4)^2 b}{16k(\beta^2-1)} \cdot \Theta_0 .$$

From the expression for u_ϱ and the equation

$$\sigma_{\varrho\varrho} - \sigma_{\phi\phi} = 2\varrho \frac{\partial}{\partial \varrho}(\varrho^{-1} u_\varrho) \tag{4.6.19}$$

we deduce that the difference of the principal stresses is

$$\begin{aligned} \sigma_{\varrho\varrho} - \sigma_{\phi\phi} = 2\Theta\{&(k-1-z)[\varrho^2 + (k-1-z)^2]^{-3/2} - \\ &-(k+1+z)[\varrho^2+(k+1+z)^2]^{-3/2} + \\ &+2(k-1-z)\varrho^{-2}[\varrho^2+(k-1-z)^2]^{-1/2} - \\ &-2(k+1+z)\varrho^{-2}[\varrho^2+(k+1+z)^2]^{-1/2}\}. \end{aligned} \tag{4.6.20}$$

Figure 3 which is taken from Sneddon and Lockett (1960a) shows the surface distribution of temperature in the case $k = 3$, together with the sections (by a plane ϑ = constant) of the isochromatic surfaces $|\sigma_{\varrho\varrho} - \sigma_{\phi\phi}|$ = *constant*. If we had chosen a higher value of the parameter k we should have obtained a surface temperature distribution which was less highly concentrated in the neighbourhood of the point $\varrho = 0$; on the other hand, if we had chosen a smaller value of k, such as $k = 2.1$, the curve would have been concentrated into a small

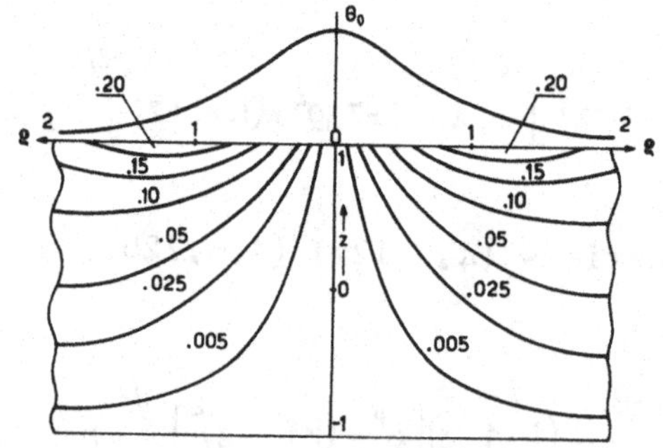

Fig. 3 Section by a plane θ = *constant* of the surfaces $|\sigma_\varrho - \sigma_\theta|$ = *constant* in a thick plate. The numbers refer to the values of $8k(\beta^2-1)(\sigma_\theta - \sigma_\varrho)/b\Theta_0(k^2-4)^2$.

band surrounding the origin.

If in equation (4.6.13) we take $f(\varrho)$ to be given by equation (4.6.17), then substituting from (4.6.18) into the set of equations (4.6.16) we obtain the displacement field

$$u_\varrho = U\varrho^{-1}\left[(k+1-z)r_3^{-1}+(k+1+z)r_2^{-1}-(k-1-z)r_1^{-1}-(k-1+z)r_2^{-1}\right] \tag{4.6.21}$$

$$u_z = U\left[r_1^{-1}-r_2^{-1}-r_3^{-1}+r_4^{-1}\right] \tag{4.6.22}$$

in which U is the constant defined by the equation

$$U = \frac{(k^2-4)^2 b}{32(\beta^2-1)k}\Theta_0$$

and the r_j ($j = 1,2,3,4$) are defined by the equations

$$r_1^2 = \varrho^2+(k-1-z)^2, \quad r_2^2 = \varrho^2+(k-1+z)^2$$

$$r_3^2 = \varrho^2+(k+1-z)^2, \quad r_4^2 = \varrho^2+(k+1+z)^2.$$

The difference of the principal stresses can again be calculated using equations (4.6.19) and (4.6.21) and the isochromatic surfaces drawn. Figures 4 and 5 which are taken from

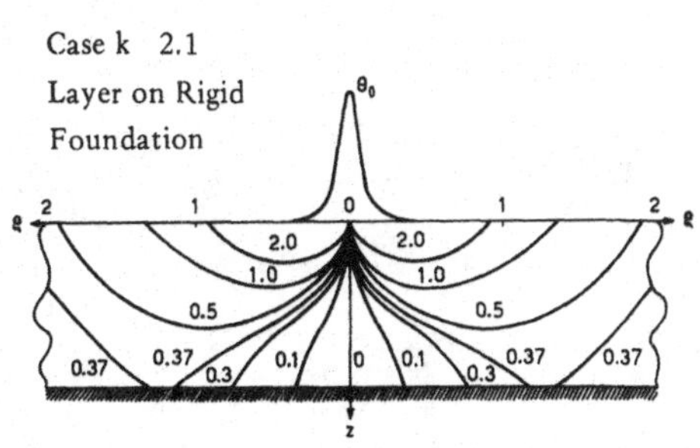

Fig. 4

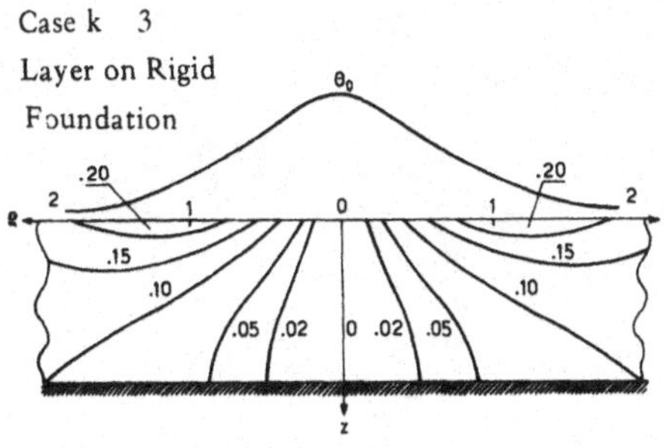

Fig. 5

Sneddon and Lockett (1960b) show the results in the two cases $K = 2.1$, $K = 3$ respectively.

Equations (4.6.21), (4.6.22) enable us to calculate the thermal stress in an elastic layer *resting* on a rigid foundation. The analysis has to be modified in the case in which the layer is *bounded* to the foundation; this problem has been considered recently by Dhaliwal (1971).

APPENDIX

INTEGRAL TRANSFORMS (*)

1. Fourier Transforms

We define the operators $\mathfrak{F}$, $\mathfrak{F}_c$ and $\mathfrak{F}_s$ by the equations

$$\mathfrak{F}[f(x);\xi] = \frac{1}{\sqrt{(2\pi)}}\int_{-\infty}^{\infty} f(x)\exp(ix\xi)dx$$

$$\mathfrak{F}_c[f(x);\xi] = \frac{2}{\pi}\int_{0}^{\infty} f(x)\cos(x\xi)dx$$

$$\mathfrak{F}_s[f(x);\xi] = \frac{2}{\pi}\int_{0}^{\infty} f(x)\sin(x\xi)dx$$

respectively. If we introduce the operator $\mathfrak{F}^*$ through the equation

$$\mathfrak{F}^*[f(x);\xi] = \mathfrak{F}[f(x);-\xi]$$

the *Fourier inversion theorem* may be written in the form

$$\hat{f}(\xi) = \mathfrak{F}[f(x);\xi] \Longrightarrow f(x) = \mathfrak{F}^*[f(x);\xi]$$

or alternatively as

$$\mathfrak{F}^{-1} = \mathfrak{F}^*.$$

(*) A useful general reference is Sneddon, 1972.

In a similar way we can write the *Fourier cosine inversion theorem* as

$$\mathfrak{F}_c^{-1} = \mathfrak{F}_c$$

and the *Fourier sine inversion theorem* as

$$\mathfrak{F}_s^{-1} = \mathfrak{F}_s \, .$$

If $f(x)$ is continuous for all real values of x we have

$$\mathfrak{F}[f'(x);\xi] = -i\xi\hat{f}(\xi) \tag{A.1}$$

but if $f(x)$ is piecewise continuous with n points of (finite) discontinuity at $a_1,\dots,a_n$ this formula should be replaced by

$$\mathfrak{F}[f'(x);\xi] = -i\xi\hat{f}(\xi) - (2\pi)^{-1/2}\sum_{r=0}^{n}[f]_{a_r}\exp(i\xi a_r) \tag{A.2}$$

where

$$[f]_{a_r} = f(a_r+) - f(a_r-) \, . \tag{A.3}$$

If $f(x)$ and its first m derivatives are continuous for all real values of x , equation (A.1) is generalized to

$$\mathfrak{F}[f^{(m)}(x);\xi] = (-i\xi)^m\hat{f}(\xi) \, . \tag{A.4}$$

Similarly, if we write

$$\hat{f}_c(\xi) = \mathfrak{F}_c[f(x);\xi] \, , \quad \hat{f}_s(\xi) = \mathfrak{F}_s[f(x);\xi] \, ,$$

we have the formulae

$$\mathfrak{F}_c[f'(t);\xi] = \xi \hat{f}_s(\xi) - (2/\pi)^{1/2} f(0) \tag{A.5}$$

$$\mathfrak{F}_s[f'(t);\xi] = -\xi \hat{f}_c(\xi) \tag{A.6}$$

from which we easily deduce that

$$\mathfrak{F}_c[f''(t);\xi] = -\xi^2 \hat{f}_s(\xi) - (2/\pi)^{1/2} f'(0) \tag{A.7}$$

$$\mathfrak{F}_s[f''(t);\xi] = -\xi^2 \hat{f}_s(\xi) + (2/\pi)^{1/2} \xi f(0). \tag{A.8}$$

Similarly if we write

$$\hat{f}_c(\xi,y) = \mathfrak{F}_c[f(x,y);x \rightarrow \xi],$$

$$\hat{f}_s(\xi,y) = \mathfrak{F}_s[f(x,y);x \rightarrow \xi], \quad f_r(y) = \sqrt{\frac{2}{\pi}}\left[\frac{\partial^r f}{\partial x^r}\right]_{x=0}$$

$$\Delta_2 = \frac{\partial^2}{\partial x^2} + \frac{\partial^2}{\partial y^2}, \quad D = \frac{\partial}{\partial y},$$

we have the formulae

$$\mathfrak{F}_c[\Delta_2 f(x,y);x \rightarrow \xi] = (D^2 - \xi^2)\hat{f}_c(\xi,y) - f_1(y) \tag{A.9}$$

$$\mathfrak{F}_s[\Delta_2 f(x,y);x \rightarrow \xi] = (D^2 - \xi^2)\hat{f}_s(\xi,y) + \xi f_0(y) \tag{A.10}$$

$$\mathfrak{F}_c[\Delta_2^2 f(x,y);x \rightarrow \xi] = (D^2 - \xi^2)^2 \hat{f}_c(\xi,y) + (\xi^2 - 2D^2) f_1(y) - f_3(y), \tag{A.11}$$

$$\mathfrak{F}_s[\Delta_2^2 f(x,y);x \rightarrow \xi] = (D^2 - \xi^2)^2 \hat{f}_s(\xi,y) - \xi(\xi^2 + 2D^2) f_0(y) + \xi f_2(y). \tag{A.12}$$

If we define the *convolution* of two functions f and g by

$$f \cdot g = (2\pi)^{-1/2} \int_{-\infty}^{\infty} f(x-u) g(u) du$$

we have the result

$$\mathfrak{F}[f \cdot g ; \xi] = \hat{f}(\xi) \hat{g}(\xi) \tag{A.13}$$

of which a special case is

$$\int_{-\infty}^{\infty} |\hat{f}(\xi)|^2 d\xi = \int_{-\infty}^{\infty} |f(t)|^2 dt . \tag{A.14}$$

Similarly we have the formulae

$$\int_{0}^{\infty} |\hat{f}_c(\xi)|^2 d\xi = \int_{0}^{\infty} |f(t)|^2 dt \tag{A.15}$$

$$\int_{0}^{\infty} |\hat{f}_s(\xi)|^2 d\xi = \int_{0}^{\infty} |f(t)|^2 dt . \tag{A.16}$$

Equations (A.14), (A.15) and (A.16) are called *Parseval's relations for Fourier transforms.*

If we have a function of n independent variables $x_1, x_2, \ldots, x_n$ we define the Fourier transform $f_{(n)}(\xi_1, \xi_2, \ldots, \xi_n)$ by the equation

$$\hat{f}_{(n)}(\vec{\xi}) = \mathfrak{F}_{(n)}[f(\vec{x}); \vec{x} \to \vec{\xi}] = (2\pi)^{-1/2n} \int_{E_n} f(\vec{x}) \exp\{i(\vec{\xi} \cdot \vec{x})\} d\vec{x} \tag{A.17}$$

where E_n denotes the n-dimensional (real) Euclidean space and

$$(\vec{x} \cdot \vec{\xi}) = x_1 \xi_1 + x_2 \xi_2 + \ldots + x_n \xi_n .$$

Defining the convolution by

$$(f \cdot g)(\vec{x}) = (2\pi)^{-1/2n} \int_{E_n} f(\vec{x}-\vec{u})g(\vec{u})d\vec{u}$$

we have the convolution theorem

$$\mathfrak{F}_{(n)}[(f \cdot g)(\vec{x}); \vec{\xi}] = \hat{f}_{(n)}(\vec{\xi})\hat{g}_{(n)}(\vec{\xi}) . \tag{A.18}$$

Corresponding to equations (A.11) and (A.12) we have

$$\mathfrak{F}_{(n)}[\Delta_n^r f(x); \xi] = (-1)^r \xi^{2r} \hat{f}_n(\xi) \tag{A.19}$$

where Δ_n^r is defined by the equations

$$\Delta_n = \frac{\partial^2}{\partial x_1^2} + \frac{\partial^2}{\partial x_2^2} + \dots + \frac{\partial^2}{\partial x_n^2}$$

and

$$\Delta_n^r = \Delta_n \Delta_n^{r-1}$$

and ξ^2 by the equation

$$\xi^2 = \xi_1^2 + \xi_2^2 + \dots + \xi_n^2 .$$

In applications we often have to find the Fourier transform of a function $f(r)$ which is a function of

$$r = |\vec{x}| = \sqrt{(x_1^2 + x_2^2 + \dots + x_n^2)}$$

only. It turns out that

$$\mathfrak{F}_{(n)}[f(r); \xi] = \xi^{-\nu} \mathcal{H}_\nu[r^\nu f(r); \xi], \quad \left(\nu = \frac{1}{2}n - 1\right), \tag{A.20}$$

where the operator $\mathcal{H}_\nu$ is defined by the equation

$$\mathcal{H}_\nu[f(x);\xi] = \int_0^\infty x f(x) J_\nu(\xi x)\,dx$$

and $\mathcal{H}_\nu f$ is called the *Hankel transform of order* ν of the function f . In particular, if $\varrho^2 = x_1^2 + x_2^2$, $\lambda^2 = \xi_1^2 + \xi_2^2$,

$$\mathcal{F}_{(2)}[f(\varrho);(\xi_1,\xi_2)] = \mathcal{H}_0[f(\varrho);\lambda] , \tag{A.21}$$

and if $r^2 = x_1^2 + x_2^2 + x_3^2$, $\mu^2 = \xi_1^2 + \xi_2^2 + \xi_3^2$,

$$\mathcal{F}_{(3)}[f(r);(\xi_1,\xi_2,\xi_3)] = \mu^{-1}\mathcal{F}_s[rf(r);\mu] . \tag{A.22}$$

2. Laplace Transforms

We define the *Laplace transform* of a function $f(x)$ on the positive real line by the equation

$$\bar{f}(p) = \mathcal{L}[f(x),p] = \int_0^\infty f(x)e^{-px}\,dx . \tag{A.23}$$

If $\bar{f}(p)$ is an analytic function of the complex variable p and is of order $O(p^{-k})$ in some half-plane $\mathrm{Re}\,p > \gamma$, where γ and k are real constants ($k > 1$) , then as $\omega \to \infty$

$$\frac{1}{2\pi i}\int_{c+i\omega}^{c+i\omega} e^{px}\bar{f}(p)\,dp , \quad (c > \gamma)$$

converges to a function $f(c)$ which is independent of c and whose Laplace transform is $\bar{f}(p)$. We shall refer to this result as the *Laplace inversion theorem*.

We also write (A.23) in the inverse form

$$f(x) = \mathcal{L}^{-1}[\bar{f}(p);x].$$

Corresponding to equation (A.2) we have the relation

$$\mathcal{L}[f'(x);p] = p\bar{f}(p) - f(0) - \sum_{r=1}^{\infty} [f]_{a_r} \exp(-pa_r) \qquad (A.24)$$

which, in the case in which f is continuous reduces to

$$\mathcal{L}[f'(x);p] = p\bar{f}(p) - f(0). \qquad (A.25)$$

Similarly if $f \in C^{n-1}(R^+)$, we have

$$\mathcal{L}[f^{(n)}(x);p] = p^n \bar{f}(p) - \sum_{r=0}^{n-1} p^{n-r-1} f^{(r)}(0). \qquad (A.26)$$

The convolution integral appropriate to the Laplace transform is

$$(f^* g)(x) = \int_0^x f(x-u)g(u)du$$

which has the property

$$\mathcal{L}[(f^* g)(x);p] = \bar{f}(p)\bar{g}(p). \qquad (A.27)$$

A straightforward application of the Laplace transform shows that the integral equation

$$\int_0^x \frac{f(t)dt}{(x^2-t^2)^a} = g(x), \quad x > 0, \quad 0 < a < 1$$

has solution

$$f(t) = \frac{2\sin(\pi a)}{\pi} \frac{d}{dt} \int_0^t \frac{xg(x)dx}{(t^2-x^2)^{1-a}} , \qquad x > 0 ,$$

and that the integral equation

$$\int_x^\infty \frac{f(t)dt}{(t^2-x^2)^a} = g(x), \qquad x > 0, \qquad 0 < a < 1$$

has solution

$$f(t) = -\frac{2\sin(\pi a)}{\pi} \frac{d}{dt} \int_t^\infty \frac{xg(x)dx}{(x^2-t^2)^{1-a}} .$$

We may write these results in the case $a = \frac{1}{2}$ in the form of inversion theorems for the *Abel transforms* $\mathcal{A}_1, \mathcal{A}_2$ defined by the equations

$$\hat{f}_1(x) \equiv \mathcal{A}_1[f(t);x] = \sqrt{\frac{2}{\pi}} \int_0^x \frac{f(t)dt}{\sqrt{(x^2-t^2)}} , \qquad x > 0 \tag{A.28}$$

$$\hat{f}_2(x) \equiv \mathcal{A}_2[f(t);x] = \sqrt{\frac{2}{\pi}} \int_x^\infty \frac{f(t)dt}{\sqrt{(t^2-x^2)}} , \qquad x > 0 . \tag{A.29}$$

The appropriate inversion formulae are

$$f(t) = \mathcal{A}_1^{-1}[\hat{f}_1(x);t] \equiv D_t \mathcal{A}_1[x\hat{f}_1(x);t] \tag{A.30}$$

$$f(t) = \mathcal{A}_2^{-1}[\hat{f}_2(x);t] \equiv -D_t \mathcal{A}_2[x\hat{f}_2(x);t] \tag{A.31}$$

with $D_t = d/dt$.

3. Hankel Transforms

The Hankel transform of order ν, $\mathcal{H}_\nu f$, of a function f was defined in the previous section.

From the recurrence formulae for the Bessel functions of the first kind we deduce readily that

$$\mathcal{H}_\nu\left[\rho^{\nu-1}\frac{\partial}{\partial\rho}\{\rho^{1-\nu}f(\rho)\};\xi\right] = -\xi\mathcal{H}_{\nu-1}[f(\rho);\xi] \tag{A.32}$$

$$\mathcal{H}_\nu\left[\rho^{-\nu-1}\frac{\partial}{\partial\rho}\{\rho^{\nu}f(\rho)\};\xi\right] = \xi\mathcal{H}_{\nu+1}[f(\rho);\xi] \tag{A.33}$$

and from these deduce that

$$\mathcal{H}_\nu[\mathcal{B}_\nu f(\rho);\xi] = -\xi^2\mathcal{H}_\nu[f(\rho);\xi] \tag{A.34}$$

where $\mathcal{B}_\nu$ denotes the differential operator

$$\frac{\partial^2}{\partial\rho^2}+\frac{1}{\rho}\frac{\partial}{\partial\rho}-\frac{\nu^2}{\rho^2}\ .$$

If Δ_3 is expressed in terms of cylindrical coordinates

$$\mathcal{H}_\nu[\Delta_3 f(\rho,z)e^{i\nu\phi};\rho\to\xi] = (D^2-\xi^2)\bar{f}_\nu(\xi,z)e^{i\nu\phi} \tag{A.35}$$

where $\bar{f}_\nu(\xi,z) = \mathcal{H}_\nu[f(\rho,z);\rho\to\xi]$ and $D = \partial/\partial z$.

In problems in which there is axial symmetry, we

may, by choosing the axis of symmetry to be the z-axis, take the Laplacian operator to be

$$\Delta_a = \mathcal{B}_0 + D^2$$

in which case the result corresponding to (A.35) is

$$\mathcal{H}_0[\Delta_a f(\varrho, z); \varrho \to \xi] = (D^2 - \xi^2)\bar{f}_0(\xi, z) \tag{A.36}$$

Putting $\nu = 1$ in (A.32) and $\nu = 0$ in (A.33) we obtain the important special cases

$$\mathcal{H}_1\left[\frac{\partial f}{\partial \varrho}; \varrho \to \xi\right] = -\xi \bar{f}_0(\xi) \tag{A.37}$$

$$\mathcal{H}_0\left[\frac{1}{\varrho} \cdot \frac{\partial f}{\partial \varrho}; \varrho \to \xi\right] = \xi \bar{f}_1(\xi) . \tag{A.38}$$

There are close connections between Hankel and Fourier transforms. For instance, we have the equations

$$\mathcal{H}_0[\hat{f}_c(\xi); \varrho] = -\mathcal{A}_2[f'(t); \varrho] \tag{A.39}$$

$$\mathcal{H}_0[\xi^{-1}\hat{f}_c(\xi); \varrho] = \mathcal{A}_1[f(t); \varrho] \tag{A.40}$$

$$\mathcal{H}_0[\hat{f}_s(\xi); \varrho] = \mathcal{A}_1[f'(t); \varrho] \tag{A.41}$$

$$\mathcal{H}_0[\xi^{-1}\hat{f}_s(\xi); \varrho] = \mathcal{A}_2[f(t); \varrho] \tag{A.42}$$

where $\hat{f}_c$ and $\hat{f}_s$ denotes respectively the Fourier cosine and sine transforms of f.

Similarly, if we denote the Hankel transform of

order 0 of f by $\bar{f}_0$ we have the equations

$$\mathcal{F}_c[\bar{f}_0(\xi);x] = \mathcal{A}_2[\varrho f(\varrho);x]$$

$$\mathcal{F}_s[\bar{f}_0(\xi);x] = \mathcal{A}_1[\varrho f(\varrho);x] .$$

Two further relations which are useful in applications are

$$\mathcal{F}_c[\mathcal{H}_0\{t^{-1}g(t)H(a-t);\xi\};x] = (2/\pi)^{1/2}H(a-x)\int_x^a \frac{g(t)dt}{\sqrt{(t^2-x^2)}} \tag{A.45}$$

$$\mathcal{F}_s[\mathcal{H}_0\{t^{-1}g(t)H(a-t);\xi\};x] = (2/\pi)^{1/2}\int_0^{\min(x,a)} \frac{g(t)dt}{\sqrt{(t^2-x^2)}} . \tag{A.46}$$

4. Mellin Transforms

The Mellin transform $f^*(s)$ of a function $f(x);x\in R^+$, is defined by

$$f^*(s) = \mathfrak{M}[f(x);s] = \int_0^\infty x^{s-1}f(x)dx \tag{A.47}$$

and is easily shown to have the properties

$$\mathfrak{M}[f^{(n)}(x);s] = (-1)^n \frac{\Gamma(s)}{\Gamma(s-n)} f^*(s-n) \tag{A.48}$$

$$\mathfrak{M}\left[\left(x\frac{d}{dx}\right)^n f(x);s\right] = (-s)^n f^*(s) \tag{A.49}$$

$$\mathfrak{M}[x^n f^{(n)}(x);s] = (-1)^n \frac{\Gamma(s+n)}{\Gamma(s)} f^*(s) . \tag{A.50}$$

If we express the two-dimensional Laplacian operator Δ_2 in terms of plane polar coordinates ϱ and ϕ we find that

(A.51) $$\mathfrak{M}[\Delta_2 f(\varrho,\phi);\varrho\to s] = [D_\phi^2+(s-2)^2]f^*(s-2,\phi)$$

where $D_\phi = \partial/\partial\phi$ and

(A.52) $$f^*(s,\phi) = \mathfrak{M}[f(\varrho,\phi);\varrho\to s].$$

Applying (A.51) twice, we obtain

(A.53) $$\mathfrak{M}[\Delta_2^2 f(\varrho,\phi);\varrho\to s] = [D^2+(s-2)^2][D^2+(s-4)^2]f^*(s-4,\phi).$$

Use is frequently made of the Mellin transforms of certain integral expressions. For instance we have the relations

(A.54) $$\mathfrak{M}\left[x^\lambda\int_0^\infty u^\mu f(xu)g(u)du;s\right] = f^*(s+\lambda)g^*(\mu+1-\lambda-s),$$

(A.55) $$\mathfrak{M}\left[x^\lambda\int_0^\infty u^\mu f(x/u)g(u)du;s\right] = f^*(s+\lambda)g^*(\lambda+\mu+1+s).$$

The case $\lambda = 0$, $\mu = -1$ of (A.55) written in inverse form is of value

(A.56) $$\mathfrak{M}^{-1}[f^*(s)g^*(s);x] = \int_0^\infty f(x/u)g(u)u^{-1}du.$$

BIBLIOGRAPHY

BIOT, M.A., 1956, *J. Appl. Phys.*, 27, 240.

BOLEY, B.A., & WEINER, H.J., 1960, *Theory of Thermal Stresses*, Wiley, New York.

BOUSSINESQ, J., 1885, *Application des Potentiels à l'Etude de l'Equilibre et du Mouvement des Solides Elastiques*, Paris.

CHADWICK, P., *Progress in Solid Mechanics*, (Vol. 1, p. 265), North Holland Pub. Co., Amsterdam.

CHADWICK, P., & SNEDDON, I.N., 1958, *J. Mech. Phys. Solids*, 6, 223.

DAS, B.R., 1968, *Int. J. Engng. Sci.*, 6, 497.
1969, *Int. J. Engng. Sci.*, 7, 667.
1971, *Int. J. Engng. Sci.*, 9, 469.

DE GROOT, S.R., 1952, *Thermodynamics of Irreversible Processes*, North Holland Pub. Co., Amsterdam.

DERESIEWICZ, H., 1957, *J. Acoust. Soc. Amer.*, 29, 204.

DHALIWAL, R.S., 1971, *J.I.M.A.*, 7, 295

DUHAMEL, J.M.C., 1837, *J. de l'Ecole Polytech.*, 15, 1.
1838, *Mem. Acad. Sci. Savants Etrangers*, 5, 440.

EASON, G., NOBLE, B. & SNEDDON, I.N., 1955, *Phil. Trans.*, A 247, 529.

EASON, G. & SNEDDON, I.N., 1958, *Proc. Roy. Soc. Edinburgh*, A 65, 143.

GATEWOOD, B.E., 1957, *Thermal Stresses*, McGraw-Hill, New York.

GEORGE, D.L., & SNEDDON, I.N., 1962, *J. Math. Mech.*, 11, 665.

GOODIER, J.N., 1937, *Phil. Mag.*, 7, 23

GREEN, A.E., & ZERNA, W., 1954, *Theoretical Elasticity*, Clarendon Press, Oxford.

JEFFREYS, H., 1930, *Proc. Cambridge Phil. Soc.* 26, 101.

KOVALENKO, A.D., 1969, *Thermoelasticity*, Wolters-Noordhoff, Gröningen.

LESSEN, M., 1965, *J. Mech. Phys. Solids*, 5, 57.
1957, *Quart. Appl. Math.*, 15, 105.

LESSEN, M., & DUKE, C.E., 1953, *Proc. 1st Midwest. Conf. Mech. Solids.*

LOCKETT, F.J., 1958, *J. Mech. Phys. Solids*, 7, 71.

LOCKETT, F.J., & SNEDDON, I.N., 1959, *Proc. Edinburgh Math. Soc.*, 11, 237.

LUR'E, A.I., 1955, *Three Dimensional Problems of the Theory of Elasticity*, (in Russian), Moscow, pp. 191-199.

McDOWELL, E.L., 1957, *Proc. 3rd. Midwest. Conf. Solid. Mech.*

MELAN, E. & PARKUS, H., 1953, *Wärmespannungen*, Springer, Wien.

MUKI, R., 1956, *Proc. Fac. Engng. Keio Univ.*, 9, 42.

MUSKHELISHVILI, N.I., 1953, *Some Basic Problems of the Mathematical Theory of Elasticity*, Noordhoff, Gröningen.

NEUMANN, F.E., 1885, *Vorlesungen über die Theorie der Elastizität der Festen Körper und des Lichtäthers*, Teubner, Leipzig.

NOWACKI, W., 1962, *Thermoelasticity*, Pergamon Press, Oxford.

PARIA, G., 1958, *Appl. Sci. Res.*, 7, 463.

PARKUS, H., 1959, *Instationäre Wärmespannungen*, Springer, Wien.

SNEDDON, I.N., 1958, *Proc. Roy. Soc. Edinburgh*, A 65, 121.
1960, *Boundary Problems in Differential Equations*, University of Wisconsin Press, Madison, Wis., pp. 231-240.
1962, *Arch. Mech. Stos.*, 14, 113.
1972, *The Use of Integral Transforms*, McGraw-Hill, New York.

SNEDDON, I.N., & BERRY, D.S., 1958, *The Classical Theory of Elasticity*, Handbuch der Physik, Bd. 6, pp. 1-124.

SNEDDON, I.N., & LOCKETT, F.J., 1960 a, *Quart. Appl. Math.*, 18, 145, 1960 b, *Annali di Mat. Pura ed Appl.*, (iv), 50, 309.

SNEDDON, I.N., & TAIT, R.J., 1961, *Problems of Continuum Mechanics*, SIAM, Philadelphia, pp. 497-512.

STERNBERG, E.,& McDOWELL, E.L., 1957, *Quart. Appl. Math.*, 14, 381.

TWEED, J., 1961, *Glasgow Math Journal*, 10, 169.

VOIGT, W., 1910, *Lehrbuch der Kristallphysik*, Teubner, Leipzig.

WATSON, G.N., 1944, *A Treatise on Bessel Functions*, Cambridge Univ. Press.

WEINER, J.H., 1957, *Quart. Appl. Math.*, 15, 102.

CONTENTS

Printed in the United States
By Bookmasters